Deep Learning for Computational Imaging

Deep Learning for Computational Imaging

REINHARD HECKEL

Great Clarendon Street, Oxford, OX2 6DP,
United Kingdom

Oxford University Press is a department of the University of Oxford.
It furthers the University's objective of excellence in research, scholarship,
and education by publishing worldwide. Oxford is a registered trade mark of
Oxford University Press in the UK and in certain other countries

Published in the United States of America by Oxford University Press
198 Madison Avenue, New York, NY 10016, United States of America

British Library Cataloguing in Publication Data

Data available

Library of Congress Control Number: 2024947079

ISBN 9780198947172
ISBN 9780198947189 (pbk.)

DOI: 10.1093/oso/9780198947172.001.0001

Printed and bound by
CPI Group (UK) Ltd, Croydon, CR0 4YY

Acknowledgements

This book would not have been possible without the support of many people in my life. Foremost, I thank my family, Marcela and Elenora, for the support and daily joy.

For valuable discussions, insights, and collaborations that shaped my understanding and perspective of the field of deep learning-based imaging and machine learning more broadly, I would like to thank (in alphabetical order) Richard Baraniuk, Helmut Bölcskei, Akshay Chaudhari, Paul Hand, Stefan Ruschke, Ludwig Schmidt, Ilan Shomorony, Mahdi Soltanolkotabi, Kannan Ramchandran, and Martin Wainwright.

I also had the pleasure of working with many outstanding students, and would like to thank them for discussions, feedback on this book, as well as help with some of the figures (in alphabetical order): Mohammad Zalbagi Darestani, Ajil Jalal, Tobit Klug, Anselm Krainovic, Kang Lin, Youssef Mansour, Dave Van Veen, Simon Wiedemann, and Yundi Zhang.

This book has been work in progress since about 2018, when I first taught a seminar course on the topic at Rice University entitled "Deep networks for inference and estimation", together with Richard Baraniuk. In 2020, at the Technical University of Munich, I started a new course called "Deep learning for inverse problems", and this book is based on the notes I wrote for this course. I would like to thank the students in this course for helpful discussions and feedback on the manuscript. I would also like to thank the Electrical and Computer Engineering Department at Rice University for providing a wonderful intellectual environment during the early stages of this work, and the Technical University of Munich for their invaluable support during the later stages.

Contents

List of Figures

1
Introduction

The focus of this book is on deep learning methods for solving inverse problems, in particular inverse problems arising in imaging. Inverse problems are about reconstructing signals (often images) from measurements with computational techniques. Inverse problems arise in medical imaging, signal processing, seismology, astronomy, and in many other fields.

The history of inverse problems in imaging is strongly intertwined with the availability of compute, which is illustrated by the history of computed tomography, a medical imaging technique for imaging body parts with X-rays.

In 1895, Wilhelm Röntgen discovered X-rays, and in the same year he produced the first medical X-ray of the hand of his wife, see Figure 1.2(a). This earned him the first Nobel prize in Physics. X-ray imaging, still used frequently in clinical practice today, generates an image by passing an X-ray beam through a body part and collecting the rays on a film; see Figure 1.1 for an illustration. X-ray imaging does not require computational techniques to generate an image. However, with X-ray imaging, information about the body part get lost since an image generated by an X-ray is a projection of a three-dimensional body part onto a two-dimensional image.

In order to reconstruct a three-dimensional object from projections onto a two-dimensional plane, computational techniques are required. Computed tomography is a medical imaging technique based on sending X-rays through a body part at different angles and collecting the measurements. From those measurements, a 3D reconstruction of the body is produced with computational techniques; see Figure 1.2(b) for an example reconstruction. Computed tomography offers a much higher level of detail than X-ray imaging.

The idea to use computational techniques for imaging is relatively old; in fact computational techniques for image reconstruction even predate the availability of computers. An early breakthrough in computed tomography was achieved by Gabriel Frank, a German engineer. In 1940, before computers existed, Dr Frank filed a patent called "Verfahren zur Herstellung von Körperschittbildern mittles Röntgenstrahlen (Method for obtaining cross-sections of bodies by means of X-rays)". The patent describes an intricate mechanical machine for running the back projection image reconstruction algorithm for computing an image from computed tomography measurements. See Figure 1.2(c) from the patent illustrating the intricate mechanical machine. Variants of this algorithm are still used in modern computed tomography scanners today, but run in silica on a modern computer. In

Deep Learning for Computational Imaging. Reinhard Heckel, Oxford University Press. © Reinhard Heckel (2025).
DOI: 10.1093/oso/9780198947172.003.0001

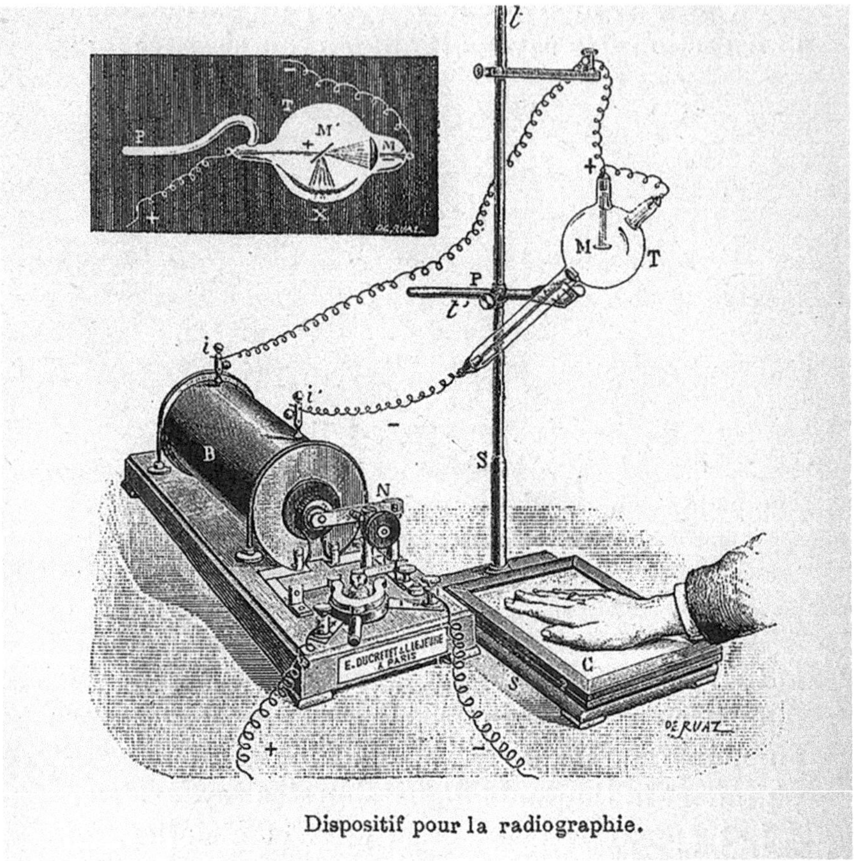

Figure 1.1 Illustration of the principle of X-ray imaging: An X-ray beam passes through a hand and is recorded on the plate beneath the hand. Attribution: René Hourdry, Travail personnel, CC BY-SA 4.0.

1971, Godfrey Hounsfield and Jamie Ambrose imaged a patient for the first time with a computed tomography scanner using computational reconstruction in a hospital in London.

Thus, as a research area, inverse problems are not especially new. Computed tomography was followed by the invention of magnetic resonance imaging in 1977, which also enables imaging anatomy through reconstructing an image with a computer, but without using potentially harmful radiation from X-rays.

What is new—and very exciting—is the use of deep learning techniques in solving important inverse problems faster and better. This development is driven by the availability of compute and image data required to work with neural networks. Until recently, most algorithms for solving inverse problems in the imaging sciences and beyond were based on static signal models derived from

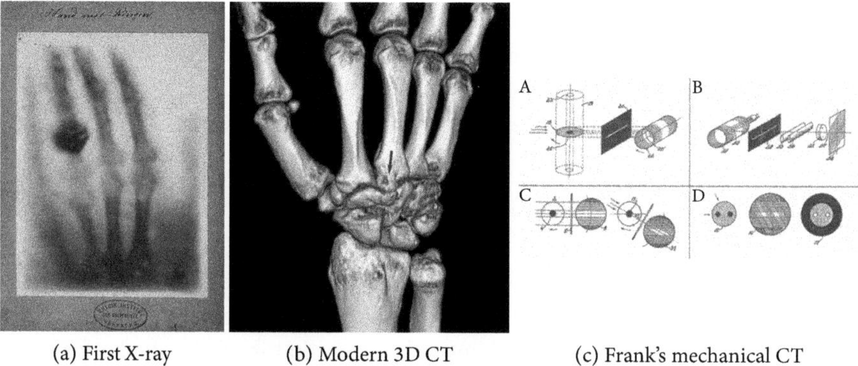

(a) First X-ray (b) Modern 3D CT (c) Frank's mechanical CT

Figure 1.2 A brief historical account of medical imaging with X-rays: (a) shows the first X-ray image produced by Röntgen in 1895. (b) shows a modern 3D reconstruction of a hand from a computed tomography scan, which is based on measurements taken with X-rays. (c) Computed tomography is older than computers itself. In his 1940 patent, Dr Gabriel Frank describes a mechanical device for computed tomography. (Copyright: (a) is a public domain image; (b) is licensed under CC BY-SA 3.0).

physics or intuition, such as wavelets or sparse representations. Much of current (2024) clinical practice in, for example, magnetic resonance imaging is based on algorithms relying on those classical, handcrafted, signal models.

Today, the best performing methods for image reconstruction and sensing problems are based on deep learning, which learn various elements of the method including (i) signal representations, (ii) parameters of iterative algorithms, (iii) regularizers, and (iv) entire inverse functions. Motivated by those success stories, researchers and practitioners are redesigning traditional imaging and sensing systems.

Deep learning-based signal reconstruction methods are already starting to be used in important medical, consumer, and scientific imaging technologies. For example, in 2019, General Electric introduced TrueFidelity, a deep learning product for computed tomography image reconstruction; in 2020, Siemens Healthineers introduced DeepResolve, a deep-learning-based reconstruction technology for denoising, and for improving image quality and/or decreasing scan times. By 2023, all big MRI vendors (Siemens, GE, Philipps, Canon Medical Systems) offered products that reconstruct images with neural networks. As a second example from consumer imaging, in the 2019 Apple reveal of the new iPhone, Apple introduced a feature called deep fusion that uses machine learning techniques to enhance images shot with a smartphone. In 2023, the Google Pixel and Apple iPhone both offered night modes that enable to reconstruct a sharp image enhanced by neural network from images taken at difficult low-light conditions.

This book gives a graduate/master level introduction to deep learning-based imaging. The book first introduces classical approaches to solving inverse problems and then aims to explain the recent advances of deep neural-network-based approaches for solving inverse problems in the imaging sciences.

1.1 What is an inverse problem?

An inverse problem has the following form: We are interested in a signal (which can be an image or a set of values) but we only have a measurement or observation of the signal obtained through applying a function that is often called a forward map. The inverse problem is to reconstruct the signal from the measurement as well as possible. See Figure 1.3 for an illustration.

Many important inverse problems are imaging problems, i.e., the signal of interest is an image or a volume (i.e., a 3D image). We focus on image reconstruction problems in this book, but the ideas and principles can also be applied to inverse problems in signal processing, communications, and beyond. Here are a few examples of inverse problems in imaging.

1.1.1 Image denoising

The perhaps most fundamental inverse problem in imaging is denoising. The goal of image denoising is to obtain an estimate of a clean image from a noisy image. Denoising problems are prevalent because essentially every imaging technology introduces noise. Even a digital camera introduces noise since the image sensor of a camera introduces noise. While the noise in modern cameras is often small, it is visible even in bright-light conditions, see Figure 1.4.

In the language of inverse problems, the forward model for denoising introduces noise, i.e., the measurement of the signal $\mathbf{x} \in \mathbb{R}^n$ is given by

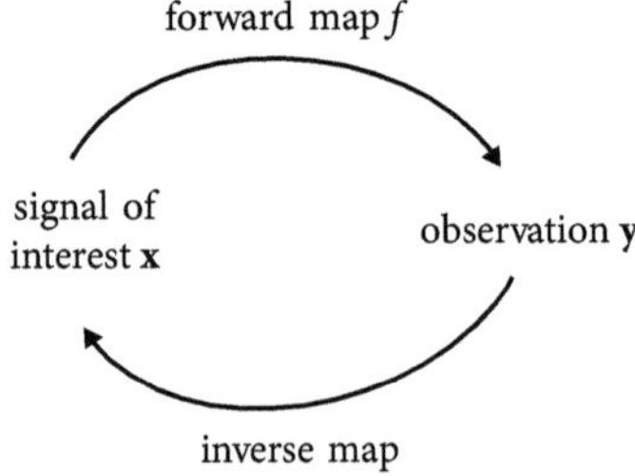

Figure 1.3 **Left**: The goal of an image recovery problem or more generally, an inverse problem, is to reconstruct a signal of interest, $\mathbf{x}$, from a measurement $\mathbf{y} = f(\mathbf{x}) + \text{noise}$. This problem occurs in any imaging system, concrete examples are camera denoising and magnetic resonance imaging (MRI).

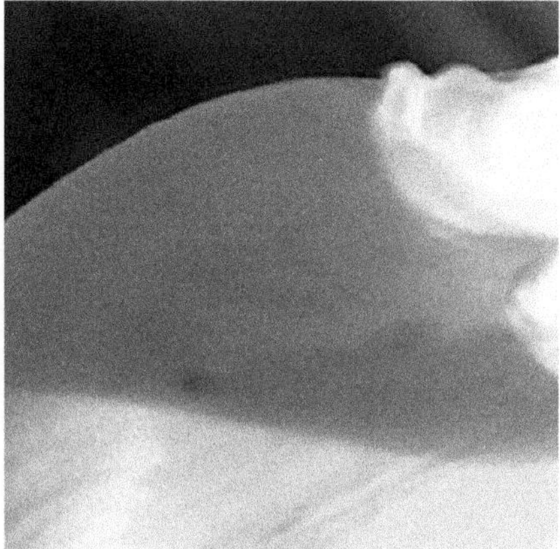

Figure 1.4 A photograph shot in raw format with a 12 megapixel smartphone camera on a bright day. Notice the small dots in the zoomed-in version on the right; this is camera noise originating from the camera sensor.

$$\mathbf{y} = \mathbf{x} + \mathbf{e},$$

where $\mathbf{e}$ is noise. Here, we model the signal $\mathbf{x} \in \mathbb{R}^n$ as a real-valued vector, where n is the number of pixel coordinates for an image, and the number of voxel coordinates for a volume or 3D image.

A very popular noise model is random Gaussian noise, added independently to each pixel of the image. Noise is a part of any imaging system and can be introduced by the sensor, and by signal processing steps, such as quantization or lossy compression. For modelling noise introduced by quantization and lossy compression, non-Gaussian and even pixel or image-dependent noise models are more appropriate than assuming Gaussian noise, but for many situations, assuming Gaussian noise is a good starting point.

Due to the randomness of the noise, it is in general impossible to exactly invert the forward map, since inverting the forward map amounts to exactly recovering the image from its noisy observation. However, as we will see later, it is possible to remove a substantial part of noise from natural images, i.e., it is possible to approximately invert the forward model.

1.1.2 Magnetic resonance imaging

Magnetic resonance imaging (MRI) is an important medical imaging technique that is used to image the anatomy of a person, and to detect pathologies such as injuries in a knee or tumours in other body parts. MRI is very popular since it is non-invasive and does not use potentially harmful radiation.

Magnetic resonance imaging is also a large active research area. The meeting of the International Society of Magnetic Resonance in Medicine (ISMRM), the largest scientific meeting focused on magnetic resonance, had over 7,000 participants in 2019, and this meeting did not even cover imaging techniques such as computed tomography or ultrasound. To put this in perspective, the International Conference on Machine Learning (ICML), one of the largest machine learning conferences, had 6,000 participants in 2019.

In a nutshell, an MRI scanner forms a strong magnetic field around the body part to be imaged. Most of the human body is made of water molecules consisting of hydrogen and oxygen atoms. The centre of a hydrogen atom contains a proton, which is sensitive to magnetic fields. Due to the magnetic field of the MRI scanner, the protons line up in the same direction. The scanner sends pulses of radio waves to certain areas of the body, which are partly absorbed by the protons and temporarily put the protons out of alignment. When the radio waves are turned off, the protons re-align which in turn emits radio signals that are picked up by receiver coils of the scanner. The resulting signal is called the magnetic-resonance (MR)-signal. From those received signals, it is possible to infer the location of the protons in the body, and in particular it enables to distinguish between different types of tissue in the body, because the protons in different tissues re-align at different speeds, thus producing distinct signals.

The goal from an algorithmic perspective is to generate an image from the MR signal measured by the MRI scanner. The MR signal typically has the form of a 2D array consisting of complex values. The complex-valued signal can be transformed into an image via a 2D Fourier transform. The resulting image is usually presented to the radiologist as a magnitude image.

In the language of inverse problems, the forward model is

$$\mathbf{y} = \mathbf{F}\mathbf{x},$$

where $\mathbf{x}$ is the complex image and $\mathbf{y}$ is the measurement (i.e., the MR signal). Since the forward model is the Fourier transform $\mathbf{F}$, it is invertible, and we can solve this inverse problem perfectly (up to inherent noise, that we ignore for simplicity).

A major challenge in MRI is that a scan takes very long—up to 90 minutes. Driving this scan duration are the iterations of excitation of the protons and detection of the protons. The scan time can be reduced by a factor of η by reducing the number of measurements taken by a factor of η, which results in the measurement model

$$\mathbf{y} = \mathbf{M}\mathbf{F}\mathbf{x},$$

where $\mathbf{M} \in \mathbb{R}^{n/\eta \times n}$ is a mask that selects a few rows of the vector $\mathbf{F}\mathbf{x}$ only. The forward map $\mathbf{M}\mathbf{F}$ is linear, but has fewer rows than columns. Thus, without any assumptions on the image $\mathbf{x}$, it is impossible to invert the forward model. We will see in this book how it is possible to nevertheless invert this forward model in practice by making prior assumptions about the image $\mathbf{x}$.

In practice, MRI measurements are not noiseless, thus we typically assume we are given a noisy measurement $\mathbf{y} = \mathbf{MFx} + \mathbf{e}$.

1.1.3 Computed tomography

Computed tomography is a medical imaging technique, where multiple X-ray measurements are taken from different angles, which results in different cross-sectional images (tomography originates from the Greek word "tomos" meaning slice or section). The forward model is

$$\mathbf{y} = \mathbf{Rx},$$

where $\mathbf{R}$ is a discrete-to-discrete linear transform that implements the action of the Radon transform on the image pixels. For so-called sparse-view sampling, the number of projections in the sinogram is small in which case $\mathbf{R}$ has more columns than rows and is, similarly as for MRI, not invertible in general.

As in MRI, the measurements are corrupted by noise, thus we typically assume we are given a noisy measurement $\mathbf{y} = \mathbf{Rx} + \mathbf{e}$.

1.1.4 Computational imaging

Classical imaging is largely based on a lens projecting an object on a film or an image sensor. The image is simply read off the sensor, and computation is not needed in principle. However, the availability of powerful computers and image reconstruction algorithms opens up new ways of designing imaging devices including devices for traditional photography and scientific imaging, such as microscopy, that are traditionally based on lenses. Computational imaging describes a research field where optical systems and reconstruction algorithms are co-designed. This is called computational imaging.

DiffuserCam [Ant+18] is an example of a computational imaging system. An image taken with a single lens cannot resolve depth because a single source of light in a scene projects on a single point on a film or image sensor. In the DiffuserCam [Ant+18], the lens is exchanged with a Diffuser, i.e., a transparent object with a relatively random, non-smooth surface. As illustrated in Figure 1.5, a single point now projects onto a certain pattern, and this pattern changes for each point in the scene. Let $\mathbf{A} \in \mathbb{R}^{m \times n}$ be the matrix which contains as columns the patterns generated by each image source on a grid in a three-dimensional volume. Then

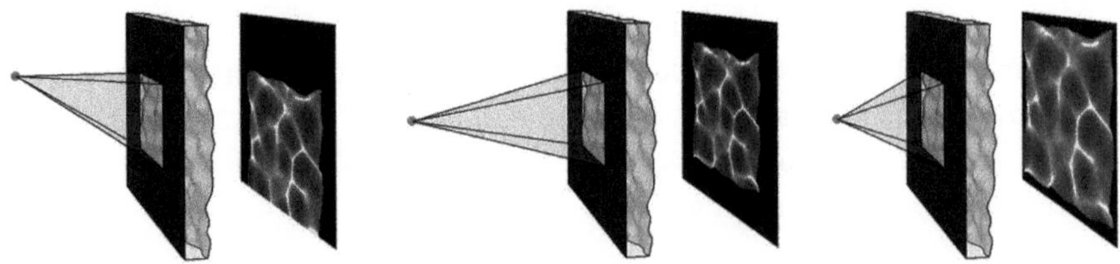

Figure 1.5 The patterns generated by differently positioned image source on the sensor of the DiffuserCam. Different image sources result in different patterns on the image sensor.

the measurement obtained by the sensor of the DiffuserCam is a super-position of all the patterns of the object $\mathbf{x}$ to be imaged:

$$\mathbf{y} = \mathbf{A}\mathbf{x}.$$

The goal is to recover the 3D image $\mathbf{x}$ from the measurements $\mathbf{y}$.

1.1.5 Cryogenic electron tomography

Cryogenic electron tomography (cryo ET) is a technique for imaging biological samples in 3D. Cryo ET is an important tool to obtain 3D models of biological samplings including cells, viruses, and proteins.

In cryo ET, a microscope collects a series of 2D projections $\mathbf{t}_{-K}, \ldots, \mathbf{t}_K$ called a tilt series of a 3D volume $\mathbf{v}$. The 2D projections and the ground truth volume are related by

$$\mathbf{F}\mathbf{t}_k = \mathbf{C}_k\mathbf{S}\mathbf{R}_k\mathbf{F}\mathbf{v} + \mathbf{F}\mathbf{e}_k, \tag{1.1}$$

where $\mathbf{F}\mathbf{t}_k$ is the 2D Fourier transform of the projected image, $\mathbf{F}\mathbf{v}$ is the 3D Fourier transform of the volume $\mathbf{v}$, the operator $\mathbf{R}_k$ spatially rotates the 3D Fourier transform, and the slice-operator $\mathbf{S}$ collects a central slice of the rotated 3D Fourier transform. Finally, $\mathbf{e}_k$ is additive noise modelled as Gaussian noise.

The goal is to reconstruct the volume $\mathbf{v}$ from the 2D projections $\mathbf{t}_{-K}, \ldots, \mathbf{t}_K$. See Figure 1.6 for an illustration.

1.1.6 Non-linear inverse problems

All example of inverse problems we discussed so far have a linear forward operator. This class of problems is called linear inverse problems. However, there are many interesting inverse problems where the forward map is non-linear. A concrete

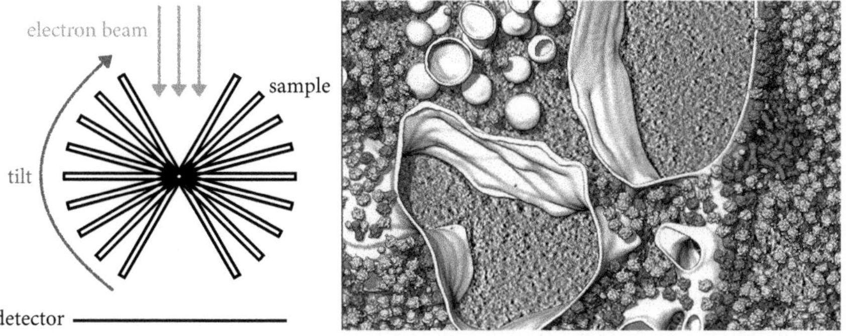

Figure 1.6 **Left**: Illustration of a cryo electron tomography. A biological sample is frozen and tilted from –60° to 60°, and for each tilt an electron beam passes through the sample and a measurement is recorded at the detector. The goal is to reconstruct the 3D volume of the sample from those 2D projections. **Right**: A visualization of the 3D volume of a sample by S. Albert et al./PNAS/CC BY 4.0.

example is coherent diffraction imaging, where the goal is to recover a complex-valued object, but we only obtain phase-less measurements. This problem is called phase-retrieval.

In most inverse problems in imaging, the forward model is known. Sometimes, however, parameters of the forward model are not known and need to be estimated during a training or tuning phase or at inference, for example calibration parameters of microscopes and sensitivity maps in magnetic resonance imaging.

1.2 Formal problem statement

To summarize the discussion from the previous section, the inverse problem considered here has the following form: Our goal is to compute an estimate $\hat{\mathbf{x}} \in \mathbb{R}^n$ of the signal $\mathbf{x} \in \mathbb{R}^n$ from a measurement

$$\mathbf{y} = f(\mathbf{x}) \in \mathbb{R}^m,$$

that is as close as possible to the original signal. Here, f is forward operator, modelling the physics of an imaging system. Typically, the forward operator is known, but there are important problems where it is not known or only partly known. We call $\mathbf{y}$ a measurement as it typically originates from a sensor or device taking measurements of a signal, or we refer to it as measurements if we refer to the entries of the vector.

Often, closeness of the original image and the estimate is measured in the mean-squared error $\frac{1}{n}||\hat{\mathbf{x}} - \mathbf{x}||_2^2$, but the performance measure is application-dependent; for example, we might measure performance in another image-quality metric such as structural similarity [Zho+04], or in a different way. We will discuss image reconstruction metrics in Chapter 14.

Throughout this book, we discuss the signal reconstruction methods primarily for two instances of inverse problems: For denoising, where the goal is to estimate the signal $\mathbf{x}$ based on a noisy measurement $\mathbf{y} = \mathbf{x} + \mathbf{e}$, and for a linear inverse problem where the goal is to estimate the signal based on a noisy measurement taken with a linear forward model $\mathbf{A}$, i.e., $\mathbf{y} = \mathbf{A}\mathbf{x} + \mathbf{e}$. The latter is fairly general and incorporates as instances accelerated MRI, computed tomography and many instances of computational imaging problems. However, the methods we discuss are usually not limited to linear inverse problems and can also be applied to inverse problems with non-linear forward maps, often with only minor adaptations.

1.3 What makes inverse problems challenging

Inverse problems are challenging to solve due to difficulties in inverting the forward map.

To illustrate that point, let us first consider the so-called compressive sensing problem, arising in MRI, CT, and the DiffuserCam examples above. In the simplest form of this problem, we obtain a measurement of an (original) image $\mathbf{x} \in \mathbb{R}^n$ as

$$\mathbf{y} = \mathbf{A}\mathbf{x},$$

where $\mathbf{A} \in \mathbb{R}^{m \times n}$ is a known measurement matrix with $m < n$. The problem of reconstructing the original image from the measurement is that there are several signals that are consistent with the measurement, and therefore reconstructing the image is, in general, impossible without making assumptions about the image. To see that there are several signals consistent with the measurement, let $\mathbf{z}$ be a vector that obeys $\mathbf{A}\mathbf{z} = 0$. Such a vector exists since the matrix $\mathbf{A}$ has more columns than rows and the columns must be linearly dependent. The vector $\mathbf{x}$ and $\mathbf{x}' = \mathbf{x} + \mathbf{z}$ map to the same measurement $\mathbf{y}$, therefore it is in general impossible to distinguish between the original image $\mathbf{x}$ and the image $\mathbf{x}'$.

Even if the forward map is invertible, it can be challenging to invert because the inverse problem is ill-conditioned. To illustrate what an ill-conditioned inverse problem is, let us consider the deconvolution problem. Let $\mathbf{x} \in \mathbb{R}^n$ be an image and let $\mathbf{y} = \mathbf{A}\mathbf{x}$ a blurred version of the image obtained by convolution with a Gaussian kernel. This convolution is implemented by the square matrix $\mathbf{A} \in \mathbb{R}^{n \times n}$. See Figure 1.7, top row, for an example image and a blurred version of the image. The matrix $\mathbf{A} \in \mathbb{R}^{n \times n}$ can be inverted which enables reconstruction of the image $\mathbf{x}$ from the blurred image $\mathbf{y}$. However, the inverse $\mathbf{A}^{-1}$ is poorly conditioned, i.e., the ratio of largest by smallest singular value is large. As a consequence, small amounts of noise added to the measurement can have a huge effect on the estimate of the signal. This is illustrated in Figure 1.7, middle row, where we add a small amount of

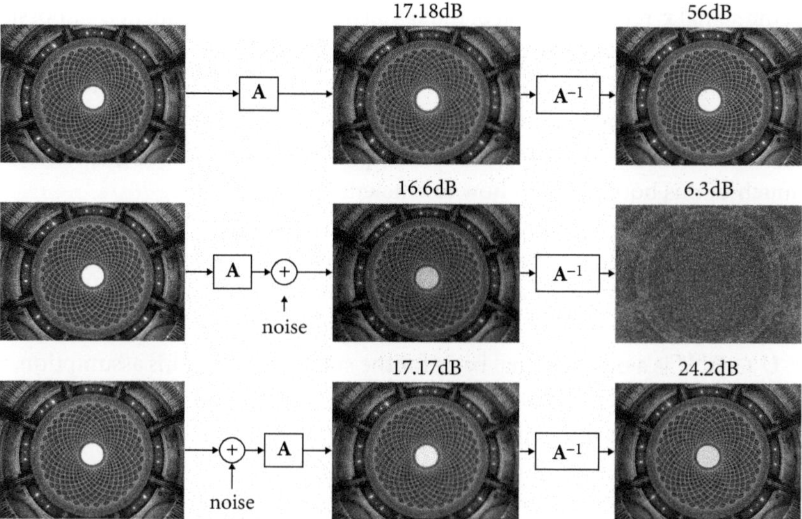

Figure 1.7 Deconvolution with a Gaussian kernel as an example of an ill-conditioned inverse problem. The PSNR relative to the original image is above each image. The linear operator (or matrix) $\mathbf{A}$ implements a convolution with a 7×7 Gaussian kernel. **Top row**: Applied to an image, the operator yields a blurry image. The inverse $\mathbf{A}^{-1}$ deconvolves the blurry image, and gives an image that is essentially perfect (56dB instead of ∞dB is due to working with floating point numbers). **Middle row**: The inverse $\mathbf{A}^{-1}$ is poorly conditioned which we can see if we add a small amount of Gaussian noise to the blurry image. Applied to the noisy measurement, the inverse $\mathbf{A}^{-1}$ generates a very noisy image (6.3dB). **Bottom row**: If the same amount of noise is added to the original image, it has a much smaller, essentially invisible effect.

noise $\mathbf{e}$ to the noise-free measurement $\mathbf{Ax}$, and apply the inverse $\mathbf{A}^{-1}$ to the noisy measurement $\mathbf{Ax} + \mathbf{e}$. It can be seen that a small amount of measurement noise has a huge effect on the reconstructed image. If the same amount of noise is added to the original image, the effect of the noise is almost imperceivable (bottom row). An inverse problem where measurement noise has an outsized effect when the forward map is directly inverted is called an ill-conditioned inverse problem.

1.4 Signal models

Signal models are central for solving inverse problems.

To see this, consider the compressive sensing problem discussed in the previous section, were we obtain a measurement of an image $\mathbf{x} \in \mathbb{R}^n$ as $\mathbf{y} = \mathbf{Ax}$, with a known measurement matrix $\mathbf{A} \in \mathbb{R}^{m \times n}$ with $m < n$. Without prior knowledge about the signal $\mathbf{x}$, reconstruction of the signal from the measurement is impossible since multiple signals are consistent with the same measurement.

What comes to our rescue here is that in most applications, we have prior knowledge of the signal to be reconstructed. This prior knowledge can be abstract and difficult to model mathematically, for example "$\mathbf{x}$ is a natural image, not noise", and much of this book is about how to efficiently model signals.

To understand how a signal model transforms an ill-posed signal model into a well-posed one, suppose that we have the prior knowledge that the signal lies in a d-dimensional subspace of $\mathbb{R}^n$, i.e., $\mathbf{x} \in \{\mathbf{x} = \mathbf{Uc}, \mathbf{c} \in \mathbb{R}^d\}$, where the dimension of the subspace is not larger than the number of measurements (i.e., $d \leq m$). Here, $\mathbf{U} \in \mathbb{R}^{n \times d}$ is a orthonormal basis for the subspace. With this assumption, we can reconstruct the signal from the measurement as follows. First, we estimate the coefficient vector $\hat{\mathbf{c}}$ from the measurement $\mathbf{y}$ as

$$\hat{\mathbf{c}} = (\mathbf{AU})^{\dagger}\mathbf{y},$$

where $(\cdot)^{\dagger}$ denotes the pseudo-inverse of a matrix. For a matrix with full column rank, $\mathbf{B}^{\dagger} = (\mathbf{B}^T\mathbf{B})^{-1}\mathbf{B}^T$ is the pseudo-inverse of the matrix. Second, we recover the signal as $\hat{\mathbf{x}} = \mathbf{U}\hat{\mathbf{c}}$. This approach provably succeeds in recovering every signal in the subspace provided that the matrix $\mathbf{AU}$ has full column rank, in which case the system of equations $\mathbf{AUc} = \mathbf{y}$ has a unique solution.

This illustrates that with low-dimensional signal models, it is possible to solve inverse problems. For many signals of interest in practice, such as images, however, linear subspaces are not a good model. In this book, we will discuss better low-dimensional signal models, such as sparse models and in particular neural-network-based signal models.

2
Solving inverse problems with optimization tasks

A solution to an inverse problem is often sought by solving an optimization problem of the form

$$\hat{\mathbf{x}} = \arg\min_{\mathbf{x}} \mathrm{DC}(\mathbf{x}, \mathbf{y}) + \lambda \mathrm{R}(\mathbf{x}), \tag{2.1}$$

where $\mathrm{DC}(\mathbf{x}, \mathbf{y})$ is a data consistency or data fidelity term that is small if the candidate signal $\mathbf{x}$ is consistent with the measurement $\mathbf{y}$, and R is a regularizer, which is small if $\mathbf{x}$ is likely, according to prior knowledge about the signal to be reconstructed. The regularization parameter $\lambda \geq 0$ trades off the data consistency and the regularizer.

The choice of the data consistency term is determined by the forward model and the noise model, and the regularizer incorporates assumptions about the signal. For example, suppose our goal is to reconstruct a signal from the measurement $\mathbf{y} = \mathbf{A}\mathbf{x} + \mathbf{e}$, where $\mathbf{e}$ is modelled by Gaussian noise. Then a natural choice for the data consistency term, motivated below, is $\mathrm{DC}(\mathbf{x}, \mathbf{y}) = \frac{1}{2}\|\mathbf{A}\mathbf{x} - \mathbf{y}\|_2^2$. Another measurement or noise model results in a different data consistency term. If we further assume that the signal $\mathbf{x}$ is sparse, then a good choice for the regularizer is the ℓ_1-norm, $\mathrm{R}(\mathbf{x}) = \|\mathbf{x}\|_1$. Sparse signals and ℓ_1-norm regularization are discussed in Chapter 4.

Optimization-based reconstruction methods work well if the forward model and noise statistics are known and if the distribution (or prior assumptions) of the signal can be well-modelled by a regularizer. In setups where this is not the case, the deep learning-based methods discussed later can improve significantly over optimization-based methods.

2.1 Bayesian estimation perspective

The optimization problem (2.1) for signal reconstruction can be motivated by taking an Bayesian estimation perspective. Suppose our goal is to estimate an unknown signal $\mathbf{x}$ given an observation $\mathbf{y}$, and that a conditional probability $p(\mathbf{x}|\mathbf{y})$ relates the two quantities.

Deep Learning for Computational Imaging. Reinhard Heckel, Oxford University Press. © Reinhard Heckel (2025).
DOI: 10.1093/oso/9780198947172.003.0002

In the context of Bayesian statistics, we assume that the signal $\mathbf{x}$ follows a distribution with density $p(\mathbf{x})$ called prior distribution and the conditional probability $p(\mathbf{x}|\mathbf{y})$ is referred to as a posterior probability, as it reflects the updated belief about the signal $\mathbf{x}$ after taking into account the measurement $\mathbf{y}$. The terms "prior" and "posterior" come from Latin and mean "before" and "coming after".

A popular estimator for the signal $\mathbf{x}$ is the maximum a posteriori probability (MAP) estimator, which chooses the signal that maximizes the posterior probability density function. In Bayesian statistics, "a posteriori", which is Latin for "from the latter" refers to probability distributions or estimates that are based on observed data. Contrary, "a priori" refers to knowledge or reasoning derived from theoretical deduction rather than observation.

With Bayes' rule, the MAP estimate is

$$\begin{aligned}
\hat{\mathbf{x}} &= \arg\max_{\mathbf{x}} p(\mathbf{x}|\mathbf{y}) \\
&\stackrel{\text{(i)}}{=} \arg\max_{\mathbf{x}} \frac{p(\mathbf{y}|\mathbf{x})p(\mathbf{x})}{p(\mathbf{y})} \\
&\stackrel{\text{(ii)}}{=} \arg\max_{\mathbf{x}} p(\mathbf{y}|\mathbf{x})p(\mathbf{x}) \\
&\stackrel{\text{(iii)}}{=} \arg\max_{\mathbf{x}} \log\left(p(\mathbf{y}|\mathbf{x})p(\mathbf{x})\right) \\
&= \arg\min_{\mathbf{x}} -\log p(\mathbf{y}|\mathbf{x}) - \log p(\mathbf{x}). \qquad (2.2)
\end{aligned}$$

Here, equality (i) is by Bayes' rule, and equality (ii) follows from $p(\mathbf{y})$ not depending on the optimization variable $\mathbf{x}$. Equality (iii) holds since the function $\log(\cdot)$ is monotonically increasing, and the last inequality holds because maximizing an objective is equivalent to minimizing the negative of the objective.

The first term in the objective of the MAP estimator (2.2), $-\log p(\mathbf{y}|\mathbf{x})$, is called the log-likelihood and models the relation between the candidate image $\mathbf{x}$ and the measurement $\mathbf{y}$.

The most common example for the log-likelihood term for inverse problems is a least-squares loss, motivated as follows. Suppose the data is generated as $\mathbf{y} = \mathbf{A}\mathbf{x}_* + \mathbf{e}$, where $\mathbf{x}_*$ is the signal of interest and $\mathbf{e}$ is zero-mean Gaussian noise with co-variance matrix $\sigma^2\mathbf{I}$. The noise $\mathbf{e}$ captures measurement noise, which is often modelled as Gaussian distribution. Under this assumption, the conditional probability $p(\mathbf{y}|\mathbf{x})$ is Gaussian with mean $\mathbf{A}\mathbf{x}$ and co-variance matrix $\sigma^2\mathbf{I}$, and hence is proportional to $p(\mathbf{y}|\mathbf{x}) \sim e^{-\frac{1}{2\sigma^2}\|\mathbf{A}\mathbf{x}-\mathbf{y}\|_2^2}$. Thus, the log-likelihood is

$$-\log p(\mathbf{y}|\mathbf{x}) = \frac{1}{2\sigma^2}\|\mathbf{A}\mathbf{x} - \mathbf{y}\|_2^2 + c,$$

where c is a constant that is irrelevant for the optimization problem.

For a digital camera in relatively bright conditions, assuming Gaussian measurement noise is accurate. In a digital camera, measurement noise arises from photon noise and other sensor-related sources of noise. Photon noise and other sensor-related sources contribute to the noise in proportions that depend on the signal level. Photon noise arises from the nature of the signal itself, not from any imperfections of the image sensor, since an image sensor measures the number of photons over a time interval, which is Poisson distributed. For large photon counts that occur in bright conditions, the Poisson distribution and thus photon noise is very well approximated by a Gaussian distribution. In bright conditions, photon noise is the dominant source of noise in a camera, and thus in bright conditions, camera noise is well approximated by Gaussian noise.

However, for other imaging systems and imaging conditions, assuming the noise to be Gaussian might not be an accurate assumption and a Poisson, Laplace, or another distribution might be a better model. A non-Gaussian noise model leads to log-likelihood different from the least-square loss, and thus to a different objective to be minimized.

The second term in the objective of the MAP estimator (2.2), $-\log p(\mathbf{x})$, is often referred to as the prior as it incorporates knowledge about the statistics of the signal $\mathbf{x}$. It is also referred to as a regularizer, as it helps to condition the optimization problem to yield good solutions for setups, where the likelihood alone does not determine a unique or stable solution. We write this term as $\lambda\sigma^2 \mathrm{R}(\mathbf{x}) = -\log p(\mathbf{x})$, where R stands for the regularizer and $\lambda \geq 0$ is a regularization parameter. With this notation, the MAP estimate (2.2) becomes the estimate (2.1).

To summarize, the formulation (2.1) encompasses a wide range of inverse problems, linear and non-linear. For a linear inverse problem with Gaussian measurement noise, the MAP estimate is

$$\hat{\mathbf{x}} = \arg\min_{\mathbf{x}} \frac{1}{2}\|\mathbf{A}\mathbf{x} - \mathbf{y}\|_2^2 + \lambda \mathrm{R}(\mathbf{x}). \tag{2.3}$$

2.2 Common regularizers

The choice of the regularizer is critical as it incorporates the prior assumption about the signal or image. Here are a few popular examples of priors for image reconstruction problems:

- ℓ_2-regularization. A classical regularizer is the ℓ_2-norm $\mathrm{R}(\mathbf{x}) = \|\mathbf{x}\|_2^2$, also known as Tikhonov regularization, named after the pioneering work on regularization by Andrey Tikhonov, and popularized in his monograph [Tik77]. The corresponding optimization problem is also known as ridge regression. Below, we discuss how deconvolution can be made more stable through

Tikhonov regularization. However, for image reconstruction, Tikhonov regularization is not a good choice as it leads to overly smooth images.

- Total-variation norm regularization. Here, $R(\mathbf{x}) = \sum_i |x_i - x_{i+1}|$, or expressed differently, the regularizer is the ℓ_1-norm of the gradient. This choice has been introduced for denoising by [ROF92] and had great success since the solution can be efficiently calculated and the method can retain sharp edges. For example, for accelerated magnetic resonance imaging, total-variation norm regularization works well [LDP07].
- Sparse regularization. Here, $R(\mathbf{x}) = \|\mathbf{D}\mathbf{x}\|_1$, where $\mathbf{D}$ is a sparsifying basis. Total-variation norm regularization is a special case of sparse regularization (obtained by setting $\mathbf{D}$ to the finite-difference operator). The wavelet basis as a sparsifying basis also works well for natural images. Sparse regularization is discussed in detail in Chapter 4.

2.3 Enhancing deconvolution stability through Tikhonov regularization

Let us consider an example of ℓ_2-regularization to demonstrate the tradeoff involved in regularization. We consider a simple 1D version of the deconvolution example from the introduction. We saw that for deconvolution, noise added to the measurement has a large effect on the reconstructed image.

Let $\mathbf{A} \in \mathbb{R}^{n \times n}$ be a matrix implementing a convolution with a 1D-Gaussian kernel (shown in Figure 2.1), and let $\mathbf{x} \in \mathbb{R}^n$ be a signal. The convolution of the kernel with the signal is $\mathbf{A}\mathbf{x}$. Suppose our goal is to estimate the signal $\mathbf{x}$ based on a noisy observation of the convolved image given by $\mathbf{y} = \mathbf{A}\mathbf{x} + \mathbf{e}$, where $\mathbf{e} \sim \mathcal{N}(0, \sigma^2\mathbf{I})$ is Gaussian noise.

The matrix $\mathbf{A}$ is is invertible, but poorly conditioned which implies that the noise has an outsized effect when we reconstruct by applying the inverse of $\mathbf{A}$. To see this, consider the least-squares estimate

$$\hat{\mathbf{x}}_{LS}(\mathbf{y}) = \arg\min_{\mathbf{x}} \|\mathbf{A}\mathbf{x} - \mathbf{y}\|_2^2,$$

which is given in closed form as $\hat{\mathbf{x}}_{LS}(\mathbf{y}) = \mathbf{A}^{-1}\mathbf{y}$. Let $\mathbf{A} = \mathbf{U}\boldsymbol{\Sigma}\mathbf{V}^T$ be the singular value decomposition of $\mathbf{A}$, and note that with $\mathbf{A}^{-1} = \mathbf{V}\boldsymbol{\Sigma}^{-1}\mathbf{U}^T$ we have

$$\begin{aligned} \hat{\mathbf{x}}_{LS}(\mathbf{y}) &= \mathbf{A}^{-1}\mathbf{y} \\ &= \mathbf{x} + \mathbf{V}\boldsymbol{\Sigma}^{-1}\mathbf{U}^T\mathbf{e}. \end{aligned}$$

If the noise $\mathbf{e}$ is zero, the least-squares estimator perfectly recovers the signal $\mathbf{x}$. However, non-zero noise has a large effect on the reconstruction since the matrix

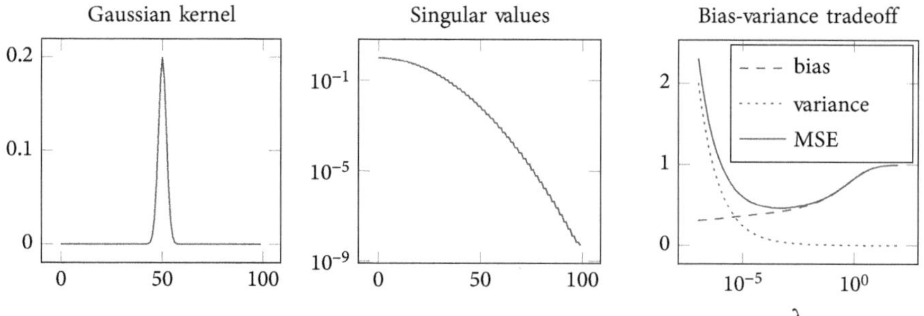

Figure 2.1 Illustration of a bias-variance tradeoff for a deconvolution problem with a Gaussian kernel. **Middle**: The associated matrix implementing the convolution with the Gaussian kernel (**left**) has very small singular values; shown are the singular values ordered from large to small. **Right**: The bias, variance (as in equation (2.4)) and the expected mean-squared-error of the regularized least-squares estimate.

$\mathbf{A}$ is poorly conditioned, meaning that the fraction of largest over the smallest singular value is very large. Thus, the inverse $\mathbf{\Sigma}^{-1} = \mathrm{diag}(1/\sigma_1, 1/\sigma_2, \ldots, 1/\sigma_n)$ contains very large values that amplify the noise. Specifically, in expectation over the noise, the mean-squared reconstruction error is

$$\begin{aligned}
\mathbb{E}\left[\|\hat{\mathbf{x}}_{LS}(\mathbf{y}) - \mathbf{x}\|_2^2\right] &= \mathbb{E}\left[\|\mathbf{V}\mathbf{\Sigma}^{-1}\mathbf{U}^T\mathbf{e}\|_2^2\right] \\
&\overset{(i)}{=} \mathbb{E}\left[\|\mathbf{\Sigma}^{-1}\mathbf{U}^T\mathbf{e}\|_2^2\right] \\
&\overset{(ii)}{=} \mathbb{E}_{\tilde{\mathbf{e}}\sim\mathcal{N}(0,\sigma^2\mathbf{I})}\left[\|\mathbf{\Sigma}^{-1}\tilde{\mathbf{e}}\|_2^2\right] \\
&= \sigma^2 \sum_{i=1}^{n} \frac{1}{\sigma_i^2},
\end{aligned}$$

where equation (i) follows from $\mathbf{V}$ being unitary, and equation (ii) follows since $\mathbf{U}^T\mathbf{e}$ is Gaussian distributed $\mathcal{N}(0, \sigma^2\mathbf{I})$, since $\mathbf{e} \sim \mathcal{N}(0, \sigma^2\mathbf{I})$ is rotationally invariant and the matrix $\mathbf{U}$ has orthonormal columns.

Next, consider the regularized least-squares estimate

$$\hat{\mathbf{x}}_\lambda(\mathbf{y}) = \arg\min_{\mathbf{x}} \|\mathbf{A}\mathbf{x} - \mathbf{y}\|_2^2 + \lambda\|\mathbf{x}\|_2^2.$$

The estimate is given in closed form by

$$\begin{aligned}
\hat{\mathbf{x}}_\lambda(\mathbf{y}) = \mathbf{V}\mathrm{diag}&\left(\frac{\sigma_1^2}{\sigma_1^2 + \lambda}, \ldots, \frac{\sigma_d^2}{\sigma_d^2 + \lambda}\right)\mathbf{V}^T\mathbf{x} \\
&+ \mathbf{V}\mathrm{diag}\left(\frac{\sigma_1}{\sigma_1^2 + \lambda}, \ldots, \frac{\sigma_d}{\sigma_d^2 + \lambda}\right)\mathbf{U}^T\mathbf{e}.
\end{aligned}$$

The expected mean-squared reconstruction error of the least-squares estimate is

$$\mathbb{E}[\|\hat{\mathbf{x}}_\lambda(\mathbf{y}) - \mathbf{x}\|_2^2] = \underbrace{\sum_{i=1}^{n} \left(1 - \frac{\sigma_i^2}{\sigma_i^2 + \lambda}\right)^2 (\mathbf{v}_i^T \mathbf{x})^2}_{\text{bias}} + \underbrace{\sigma^2 \sum_{i=1}^{n} \left(\frac{\sigma_i}{\sigma_i^2 + \lambda}\right)^2}_{\text{variance}}. \tag{2.4}$$

The first term of the expected error is a bias term that increases in the regularization parameter λ and is zero if the regularization parameter is zero ($\lambda = 0$). The second term is a variance term associated with the noise, and decreases in the regularization parameter λ, the more we regularize, the smaller the variance. Thus, the regularization parameter trades off the bias and variance terms. This is illustrated in Figure 2.1. It can be seen that with regularization we can obtain a much better estimate of the original signal $\mathbf{x}$ than without regularization, but too much regularization harms the estimate.

3
Solving optimization problems

Many problems arising in signal and image reconstruction can be formulated as an optimization problem:

$$\text{minimize} \quad L(\mathbf{x}) \quad \text{subject to} \quad \mathbf{x} \in \mathcal{C}, \tag{3.1}$$

where L is a cost function and $\mathcal{C}$ is a constraint set.

Recall, for example, the optimization based reconstruction approach discussed in Chapter 2. Another example is the ℓ_1-minimization-based reconstruction discussed in Chapter 4, where the objective is the ℓ_1-norm $L(\mathbf{x}) = \|\mathbf{x}\|_1$ and the constraint set is the data consistency $\mathcal{C} = \{\mathbf{x} : \mathbf{A}\mathbf{x} = \mathbf{y}\}$. Another example for the constraint set is approximate data consistency, such as $\|\mathbf{A}\mathbf{x} - \mathbf{y}\|_2^2 \leq \xi$.

For many common regularizers (for example ℓ_2 regularization, total-variation norm regularization, and sparse regularization), the optimization-based reconstruction approach is formulated as a convex optimization problem. Convex optimization problems have a unique global minimum making them easier to solve efficiently. In this chapter, we discuss convex optimization problems and how to solve them efficiently with iterative methods.

Moreover, training a neural network can be formulated as an optimization problem of the form (3.1). For example, consider a neural network $f_{\boldsymbol{\theta}}$ (or another machine-learning model) parameterized by $\boldsymbol{\theta}$ that maps a measurement to a (clean) signal and suppose our goal is to fit the model to a set of examples $\{(\mathbf{x}_1, \mathbf{y}_1), \ldots, (\mathbf{x}_N, \mathbf{y}_N)\}$ by minimizing the loss

$$L(\boldsymbol{\theta}) = \sum_{i=1}^{N} \|f_{\boldsymbol{\theta}}(\mathbf{y}_i) - \mathbf{x}_i\|_2^2.$$

In this example, the objective function L models how well the neural network predicts the signals $\mathbf{x}_1, \ldots, \mathbf{x}_N$ based on the corresponding measurements $\mathbf{y}_1, \ldots, \mathbf{y}_N$. Training a neural network is usually not a convex optimization problem, but the same iterative methods that we use to solve convex optimization problems are used and are effective for training neural networks.

Finally, a third motivation for studying iterative schemes for optimization problems is that some image reconstruction methods that are quite effective are based on unrolling iterative optimization schemes.

In this chapter, we start with some fundamentals of convex optimization problems, and then discuss iterative optimization algorithms, such as gradient descent,

Deep Learning for Computational Imaging. Reinhard Heckel, Oxford University Press. © Reinhard Heckel (2025).
DOI: 10.1093/oso/9780198947172.003.0003

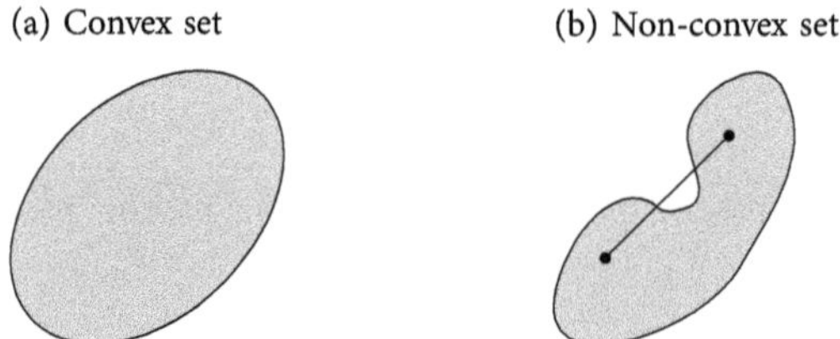

Figure 3.1 Examples of a convex and a non-convex set.

that are most important for image reconstruction and for training neural networks for image reconstruction.

3.1 Convex optimization problems

An important class of optimization problems for which we can check optimality rather easily, and which we can solve efficiently are convex optimization problems. In this section, we discuss a few fundamentals of convex optimization problems.

We say that $\mathbf{x}_*$ is a (global) solution to the optimization problem (3.1) if it is in the constraint set, i.e., $\mathbf{x}_* \in \mathcal{C}$, and if the objective is smaller than that for all other vectors in the constraint set, i.e., $L(\mathbf{x}_*) \leq L(\mathbf{x})$ for all $\mathbf{x} \in \mathcal{C}$. We say that $\mathbf{x}_*$ is a local solution to the optimization problem if in a neighbourhood $\mathcal{N}$ around $\mathbf{x}_*$, $L(\mathbf{x}_*) \leq L(\mathbf{x})$ for all $\mathbf{x} \in \mathcal{C} \cap \mathcal{N}$. Optimality of a candidate solution to an optimization problem can be very hard to check, even if the objective L is differentiable and if the constraint set is unbounded (i.e., $\mathcal{C} = \mathbb{R}^n$). An example is the ℓ_0-minimization problem discussed in the next chapter. For convex optimization problems, however, optimality is easy to check as we will see next.

To define convex optimization problems, we first need to define convex sets and functions.

Definition 1 *A convex set $\mathcal{C}$ is any set such that for all $\mathbf{x}, \mathbf{x}' \in \mathcal{C}$ and all scalars $\theta \in (0, 1)$,*

$$\theta\mathbf{x} + (1 - \theta)\mathbf{x}' \in \mathcal{C}.$$

Figure 3.1 shows examples of a convex and a non-convex set. Intersections of convex sets are convex. Important examples of convex sets are the following:

- subspaces $\{\mathbf{x} = \mathbf{U}\mathbf{a}: \mathbf{a} \in \mathbb{R}^d\}$,
- affine spaces $\{\mathbf{x} = \mathbf{U}\mathbf{a} + \mathbf{b}: \mathbf{a} \in \mathbb{R}^d\}$, and
- half-spaces $\{\mathbf{x}: \langle \mathbf{a}, \mathbf{x} \rangle \leq b\}$

are all convex sets.

Definition 2 *A convex combination of a finite set of points $\mathbf{x}_1, \dots, \mathbf{x}_N \in \mathbb{R}^n$ is a linear combination, where all coefficients are non-negative and sum up to one, i.e., $\mathbf{x}$ is a convex combination if there exist scalars $\theta_1, \dots, \theta_N$ so that*

$$\mathbf{x} = \sum_{i=1}^{N} \theta_i \mathbf{x}_i : \sum_{i=1}^{N} \theta_i = 1, \theta_i \geq 0.$$

A set $\mathcal{C}$ is convex if and only if it contains all convex combinations of its elements. For any given set $\mathcal{S}$, the convex hull of $\mathcal{S}$ is defined as the set of all convex combinations of points in $\mathcal{S}$. Intuitively, it is the smallest convex set that contains $\mathcal{S}$. As an example, consider the non-convex set $\{\mathbf{x}: \|\mathbf{x}\|_0 \leq 1, \|\mathbf{x}\|_1 \leq 1\}$. Its convex hull is the ℓ_1-norm ball $\{\mathbf{x}: \|\mathbf{x}\|_1 \leq 1\}$.

We are now ready to define a convex function.

Definition 3 *A function $L: \mathbb{R}^n \to \mathbb{R}$ is convex if for all $\mathbf{x}, \mathbf{x}' \in \mathbb{R}^n$, and all $\theta \in (0, 1)$,*

$$\theta L(\mathbf{x}) + (1 - \theta) L(\mathbf{x}') \geq L(\theta \mathbf{x} + (1 - \theta)\mathbf{x}').$$

A function L is strictly convex, if the inequality above is strict whenever $\mathbf{x} \neq \mathbf{x}'$. A function L is concave if $-L$ is convex.

Common examples of convex functions are:

- Linear functions: $L(\mathbf{x}) = \langle \mathbf{a}, \mathbf{x} \rangle + b$,
- Quadratics: $L(\mathbf{x}) = \frac{1}{2}\mathbf{x}^T\mathbf{Q}\mathbf{x} + \mathbf{b}^T\mathbf{x} + c$, where $\mathbf{Q}$ is positive semidefinite. The function L is convex if and only if $\mathbf{Q}$ is positive semidefinite, and is strictly convex if $\mathbf{Q}$ is positive definite.
- Any norm $L(\mathbf{x}) = \|\mathbf{x}\|$ is convex. This follows from the triangle inequality and homogeneity/scalability.

An important property of convex functions is that the local minima are global.

Proposition 1 *Any local minimum of a convex function $L: \mathbb{R}^n \to \mathbb{R}$ is also a global minimum.*

We now turn to first-order optimality conditions. To this end, we first state a proposition stating that a function is convex if and only if it is lower bounded by its first-order Taylor expansion at any point.

Proposition 2 *A differentiable function L is convex if and only if for all* $\mathbf{x}, \mathbf{x}' \in \mathbb{R}^n$,

$$L(\mathbf{x}') \geq L(\mathbf{x}) + (\mathbf{x}' - \mathbf{x})^T \nabla L(\mathbf{x}).$$

It is strictly convex if and only if the inequality holds strictly for all $\mathbf{x} \neq \mathbf{x}'$.

An important corollary of Proposition 2 is the following optimality condition.

Corollary 1 *If a differentiable function L is convex and* $\nabla L(\mathbf{x}_*) = 0$, *then* $\mathbf{x}_*$ *is a global minimizer of L.*

Thus, for convex and differentiable functions, we can easily verify whether a point $\mathbf{x}$ is an optimal point by computing the gradient and checking if it is zero. For a linear least-squares problem, this condition enables to derive the optimal solution. Specifically, consider the least-squares objective

$$L(\mathbf{x}) = \frac{1}{2}\|\mathbf{A}\mathbf{x} - \mathbf{y}\|_2^2.$$

Setting the gradient $\nabla L(\mathbf{x}) = \mathbf{A}^T(\mathbf{A}\mathbf{x} - \mathbf{y})$ to zero yields the condition $\mathbf{A}^T\mathbf{A}\mathbf{x}_* = \mathbf{A}^T\mathbf{y}$ for the minimizer $\mathbf{x}_*$. Provided that $\mathbf{A}^T\mathbf{A}$ is invertible, which is given if $\mathbf{A}$ has full column rank, this implies that the minimizer of the least-squares objective is $\mathbf{x}_* = (\mathbf{A}^T\mathbf{A})^{-1}\mathbf{A}^T\mathbf{y}$.

3.2 Computational aspects of optimization algorithms

Suppose we want to minimize a differentiable function $L: \mathbb{R}^n \to \mathbb{R}$ over $\mathbb{R}^n$:

$$\underset{\mathbf{x} \in \mathbb{R}^n}{\text{minimize}}\ L(\mathbf{x}).$$

As an example, consider the least-squares objective $L(\mathbf{x}) = \frac{1}{2}\|\mathbf{A}\mathbf{x} - \mathbf{y}\|_2^2$, where $\mathbf{A} \in \mathbb{R}^{m \times n}$ is a matrix with full column rank.

As discussed above, if the columns of $\mathbf{A}$ are linearly independent, for this particular example, we can simply obtain a closed-form solution as $\mathbf{x}_* = (\mathbf{A}^T\mathbf{A})^{-1}\mathbf{A}^T\mathbf{y}$. However, for most practically relevant functions L that we are interested in minimizing, we cannot compute a closed-form solution.

And even if we can compute a closed-form solution, we might not want to because computing the closed-form solution can be computationally expensive. For example, to compute the least-squares estimate via the formula $\mathbf{x}_* = (\mathbf{A}^T\mathbf{A})^{-1}\mathbf{A}^T\mathbf{y}$, we have to compute the inverse of a possibly large matrix.

We consider algorithms for finding *approximate* solutions of the minimization problem above. In practice, we are almost always content with an approximate solution, and a computer working with floating-point numbers can only give us a solution up to machine precision anyways. We assume that we can evaluate the function L at a point $\mathbf{x}$ as well as compute a derivative (or a subdifferential) $\nabla L(\mathbf{x})$.

We are interested in obtaining an approximate solution computationally efficiently. That means we are interested in an algorithm that requires a small number of floating-point operations to obtain an ϵ-accurate solution. An ϵ-accurate solution can be a vector $\mathbf{x}_k$ obeying $\|\mathbf{x}_k - \mathbf{x}_*\| \leq \epsilon$, or $L(\mathbf{x}_k) - L(\mathbf{x}_*) \leq \epsilon$. The optimization algorithms we discuss here are all first-order iterative methods, i.e., they compute a gradient and update an iterate based on the gradient. The computational cost of such an algorithm is the cost of performing an update, which is usually the cost of computing a gradient, times the number of iterations to reach an ϵ-accurate solution. Therefore it is common to characterize the number of steps required to obtain an ϵ-accurate solution, which is the convergence rate of the algorithm.

3.3 Gradient descent

Gradient descent is a simple iterative algorithm for minimizing a differentiable function $L: \mathbb{R}^n \to \mathbb{R}$ on $\mathbb{R}^n$. Starting from an initial point $\mathbf{x}_0$, gradient descent iterates:

$$\mathbf{x}_{k+1} = \mathbf{x}_k - \eta_k \nabla L(\mathbf{x}_k),$$

where η_k is a step size parameter. The idea behind the gradient descent algorithm is to make small steps in the direction that minimizes the local first-order approximation of the objective L. With an appropriate step size, gradient descent converges to a local minimum. Provided that the objective function L is convex, and with an appropriate step size, gradient descent converges to a global minimum, since for convex functions a local minimum is also a global minimum.

If the objective L is not differentiable, which is the case for many objective functions involving neural networks, we can use a subgradient instead of the gradient.

A subgradient is defined as follows. Consider a convex function $L: \mathbb{R}^n \to \mathbb{R}$, that is not necessarily differentiable. For a given $\mathbf{x} \in \mathbb{R}^n$, a subgradient is a vector $\mathbf{g} \in \mathbb{R}^n$ such that

$$L(\mathbf{x}') \geq L(\mathbf{x}) + (\mathbf{y} - \mathbf{x})^T \mathbf{g} \quad \text{for all } \mathbf{x}' \in \mathbb{R}^n. \tag{3.2}$$

In geometric terms, a subgradient $\mathbf{g}$ specifies a supporting hyperplane to the epigraph of L (i.e., the set of points lying on or above its graph) at $\mathbf{x}$. As an example, consider the function $|x|$, which is convex and not differentiable at 0. The subgradient at 0 is the interval $[-1, 1]$.

The set containing all subgradients, i.e., all vectors that satisfy inequality (3.2) is called the subdifferential and is denoted by $\partial L(\mathbf{x})$. The subdifferential is a natural generalization of the derivative of a convex function, since the gradient of a convex and differentiable function satisfies condition (3.2) with $\mathbf{g} = \nabla L(\mathbf{x})$, specifically we have $\partial L(\mathbf{x}) = \{\nabla L(\mathbf{x})\}$.

The idea of gradient descent is to move in a descent direction. However, elements of the subdifferential can be ascent directions. For example, the subdifferential $\partial L(0)$, $L(x) = |x|$ contains ascent directions. However, there always exists a descent direction in the subdifferential, and it is the element with the smallest norm in the subdifferential. This direction might be difficult to compute, however, and just finding any subgradient is easier and typically computationally cheaper. Thus, to run the subgradient method, we typically choose any subgradient of the subdifferential, and it can be shown that the subgradient method, which takes a step in any direction that lies in the subdifferential converges. The software packages that are typically used for optimizing neural networks like pytorch, lux, or tensorflow, provide us with a gradient or subgradient and we do not need to worry about which subgradient.

3.3.1 Convergence for quadratic functions

In order to understand the gradient method better, let us study convergence for a simple class of function. We consider quadratic functions parameterized with a strictly positive definite matrix $\mathbf{Q}$ and vector $\mathbf{b}$

$$L(\mathbf{x}) = \frac{1}{2}\mathbf{x}^T \mathbf{Q} \mathbf{x} - \mathbf{b}^T \mathbf{x}.$$

A closed-form solution to the minimization problem $\text{minimize}_{\mathbf{x}} L(\mathbf{x})$ is $\mathbf{x}_* = \mathbf{Q}^{-1}\mathbf{b}$.

In this section, we discuss the behaviour of gradient descent applied to this problem. We consider gradient descent with a fixed step size η:

$$\mathbf{x}_{k+1} = \mathbf{x}_k - \eta \nabla L(\mathbf{x}_k).$$

The convergence rate of gradient descent is given by the following proposition.

Proposition 3 *Gradient descent with step size $\eta = \frac{2}{M+m}$ applied to a quadratic function obeys*

$$\|\mathbf{x}_k - \mathbf{x}_*\|_2 \leq \left(\frac{1 - 1/\kappa}{1 + 1/\kappa}\right)^k \|\mathbf{x}_0 - \mathbf{x}_*\|_2.$$

Thus, the rate of convergence is linear or geometric. This rate of convergence is very fast; only a few steps can be sufficient to reach an iterate close to the optimal solution $\mathbf{x}_*$. Suppose we want to find a solution that is ϵ-close to the original solution. It follows from the proposition that we require

$$K = \log\left(\frac{1 + 1/\kappa}{1 - 1/\kappa}\right)^{-1} \log(\|\mathbf{x}_0 - \mathbf{x}_*\|_2/\epsilon)$$

many iterations to reach an ϵ-accurate solution. Due to $\log(1 + x) \approx x$ for small x, for large condition numbers we have that

$$K = O(\kappa \log(1/\epsilon)) = O(M/m \log(1/\epsilon)) \tag{3.3}$$

many iterations are sufficient for an ϵ-accurate solution.

Proof of Proposition 3: Using that the gradient is given by $\nabla L(\mathbf{x}) = \mathbf{Qx} - \mathbf{b}$ and that the optimal solution $\mathbf{x}_*$ obeys $\mathbf{Qx}_* = \mathbf{b}$, the difference of the $(k + 1)$-st iteration to the optimum is

$$\begin{aligned}
\mathbf{x}_{k+1} - \mathbf{x}_* &= \mathbf{x}_k - \eta\nabla L(\mathbf{x}_k) - \mathbf{x}_* \\
&= \mathbf{x}_k - \eta(\mathbf{Qx}_k - \mathbf{b}) - \mathbf{x}_* \\
&= \mathbf{x}_k - \eta(\mathbf{Qx}_k - \mathbf{Qx}_*) - \mathbf{x}_* \\
&= (\mathbf{I} - \eta\mathbf{Q})(\mathbf{x}_k - \mathbf{x}_*).
\end{aligned}$$

It follows that the distance to the optimal solution is bounded by

$$\|\mathbf{x}_{k+1} - \mathbf{x}_*\|_2 \leq \|\mathbf{I} - \eta\mathbf{Q}\| \|\mathbf{x}_k - \mathbf{x}_*\|_2.$$

Using that $\mathbf{I} - \eta\mathbf{Q}$ is symmetric, we get

$$\|\mathbf{I} - \eta\mathbf{Q}\| = \max(\lambda_{\max}(\mathbf{I} - \eta\mathbf{Q}), -\lambda_{\min}(\mathbf{I} - \eta\mathbf{Q})) = \max(\eta M - 1, 1 - \eta m), \tag{3.4}$$

where M and m are the largest and smallest singular values of the matrix $\mathbf{Q}$. The first equality can be checked by taking the singular value decomposition of the matrix, and for the second equality, we used that, due to $(\mathbf{I} - \eta\mathbf{Q})\mathbf{v}_i = \lambda_i(\mathbf{I} - \eta\mathbf{Q})\mathbf{v}_i$, the eigenvalues of $\mathbf{I} - \eta\mathbf{Q}$ and $\mathbf{Q}$ are related as $\lambda_i(\mathbf{I} - \eta\mathbf{Q}) = 1 - \eta\lambda_i(\mathbf{Q})$. Note the singular value decomposition and the eigenvalue decomposition of a symmetric matrix are identical. Therefore, λ can be used for singular values and eigenvalues at the same time. $\lambda_i(\mathbf{A})$ stands for the i-th eigenvalue of matrix $\mathbf{A}$.

The right-hand side of Equation (3.4) is minimized for the step size $\eta = \frac{2}{M+m}$. For this choice, $\|\mathbf{I} - \eta\mathbf{Q}\| = \frac{1-1/\kappa}{1+1/\kappa} < 1$, where $\kappa = M/m$ is the condition number of the matrix $\mathbf{Q}$. This concludes the proof of the proposition. □

3.4 Proximal gradient descent

The gradient descent algorithm introduced in the previous section is only one example of an iterative algorithm for solving a convex optimization problem, and many more exist that can be faster for a given optimization problem.

Proximal gradient descent is an algorithm that works well for an objective function that can be split into a smooth part and another part that is possibly non-differentiable but that allows for easy computation of a proximal operator. An example of such an objective function is the ℓ_1-regularized least-squares objective for sparse recovery. As we will discuss in Chapter 4, solving

$$\arg\min_{\mathbf{x}} \|\mathbf{x}\|_1 \text{subject to } \|\mathbf{A}\mathbf{x} - \mathbf{y}\|_2^2 \leq \xi, \tag{3.5}$$

provably recovers a sparse vector from an observation $\mathbf{y}$, provided the sparsity is sufficiently low, and the measurement matrix $\mathbf{A}$ satisfied certain conditions. It can be shown that the constraint problem is equivalent to the unconstrained ℓ_1-regularized least-squares problem

$$\arg\min_{\mathbf{x}} \underbrace{\frac{1}{2}\|\mathbf{A}\mathbf{x} - \mathbf{y}\|_2^2}_{f(\mathbf{x})} + \lambda\|\mathbf{x}\|_1, \tag{3.6}$$

i.e., for each ξ there is a regularization parameter λ so that the constrained optimization problem (3.5) and the ℓ_1-regularized optimization problem (3.6) yield equivalent solutions.

First, suppose we are only interested in minimizing the smooth function f. The gradient descent iterates for optimizing f with constant step size η are given by

$$\mathbf{x}_{k+1} = \mathbf{x}_k - \eta\nabla f(\mathbf{x}_k).$$

We can write those iterates in so-called proximal form as a solution to a simple optimization problem:

$$\mathbf{x}_{k+1} = \arg\min_{\mathbf{x}} \frac{1}{2}\|\mathbf{x} - (\mathbf{x}_k - \eta\nabla f(\mathbf{x}_k))\|_2^2.$$

However, our goal is to minimize the objective $L(\mathbf{x}) = f(\mathbf{x}) + \lambda\|\mathbf{x}\|_1$ instead of f, so it seems natural to solve

$$\mathbf{x}_{k+1} = \arg\min_{\mathbf{x}} \frac{1}{2}\|\mathbf{x} - (\mathbf{x}_k - \eta\nabla f(\mathbf{x}_k))\|_2^2 + \eta\lambda\|\mathbf{x}\|_1, \tag{3.7}$$

since the first term ensures that $\mathbf{x}_{k+1}$ does not deviate too far from a gradient step aimed at minimizing the function f, and the second term incentivizes to minimize the function $\lambda\|\mathbf{x}\|_1$.

Those iterates are the so-called proximal gradient descent updates. There are more formal ways to arrive at the proximal gradient descent algorithm, and it can be shown that proximal gradient descent provably converges for the loss function above.

Proximal gradient descent is popular for ℓ_1-regularized least squares, because the optimization problem in (3.7) for obtaining the iterates is separable and thus easy to solve. Specifically, since the (scaled) ℓ_1-norm $\lambda\|\mathbf{x}\|_1 = \lambda\sum_i |x_i|$ is separable, the optimization problem corresponding to the i-th coordinate is given by

$$\min_{x\in\mathbb{R}} \frac{1}{2}(x - x_i)^2 + \eta\lambda|x|, \quad x_i = [\mathbf{x}_k - \eta\nabla f(\mathbf{x}_k)]_i.$$

This coordinate-wise optimization problem has the closed-form solution

$$\arg\min_{y} \frac{1}{2}(x-y)^2 + \lambda|y| = \tau_\lambda(y), \quad \tau_\lambda(z) = \begin{cases} z+\lambda, & \text{if } z < -\lambda, \\ 0, & \text{if } z \in [-\lambda, \lambda], \\ z-\lambda, & \text{if } z > \lambda. \end{cases} \tag{3.8}$$

The operator τ_λ is called the soft-thresholding operator. The resulting algorithm is called Iterative Shrinkage-Thresholding Algorithm (ISTA).

Equation (3.8) follows by analysing the one-dimensional problem

$$\arg\min_{y} \frac{1}{2}(x-y)^2 + \lambda|y|.$$

Specifically, the optimality condition is that

$$0 \in \partial\left(\frac{1}{2}(x-y)^2 + \lambda|y|\right),$$

which is equivalent to

$$0 = y - x + \lambda\text{sign}(y).$$

3.5 The stochastic gradient method

The stochastic gradient method (SGM) is a very popular algorithm for training neural networks. We typically train a neural network f_θ with parameters $\boldsymbol{\theta}$, by minimizing a loss function of the form

$$L(\boldsymbol{\theta}) = \frac{1}{N}\sum_{i=1}^{N} \ell_i(\boldsymbol{\theta}), \tag{3.9}$$

where $\ell_i(\boldsymbol{\theta})$ is the loss associated with the i-th training example. For example, if we train a neural network in a supervised fashion to map a measurement $\mathbf{y}_i$ to a clean image $\mathbf{x}_i$ (see Chapter 6), the loss associated with the i-th training example $(\mathbf{x}_i, \mathbf{y}_i)$ is $\ell_i(\boldsymbol{\theta}) = \|\mathbf{x}_i - f_\theta(\mathbf{y}_i)\|_2^2$. To minimize the loss (3.9), we could in principle apply gradient descent. However, computation of the gradient is typically very expensive since it requires computing the gradients of the losses associated with each training example, i.e., $\nabla\ell_1(\boldsymbol{\theta}), \ldots, \nabla\ell_N(\boldsymbol{\theta})$, and typically the number of training examples N is relatively large.

Rather than relying on access to the gradient, the stochastic gradient method requires access to a stochastic gradient $G(\boldsymbol{\theta})$, which is a random variable which is equal to the gradient in expectation, i.e., $\mathbb{E}\left[G(\boldsymbol{\theta})\right] = \nabla L(\boldsymbol{\theta})$.

An example of a stochastic gradient for the loss $L(\boldsymbol{\theta})$ is the following: Choose i uniformly at random from the number of training examples $\{1, \ldots, N\}$, and let

$$G(\boldsymbol{\theta}) = \nabla\ell_i(\boldsymbol{\theta}).$$

Note that in expectation, this stochastic gradient is equal to the gradient of the loss function, i.e.,

$$\mathbb{E}\left[G(\boldsymbol{\theta})\right] = \sum_{i=1}^{N} \frac{1}{N}\nabla\ell_i(\boldsymbol{\theta}) = \nabla L(\boldsymbol{\theta}).$$

Here, we used that the loss associated with training example i is chosen with probability $1/N$. Note that the computational cost of computing the stochastic gradient is a factor N smaller than the cost of computing the gradient, if we neglect the cost of generating a random number.

Given access to a stochastic gradient, the stochastic gradient method performs the following updates, starting from an initial value $\boldsymbol{\theta}_0$:

$$\boldsymbol{\theta}_{k+1} = \boldsymbol{\theta}_k - \eta G(\boldsymbol{\theta}_k).$$

The idea is intuitive: by following the descent direction in expectation, we should get close to the optimal solution, provided we iterate long enough. However, in each individual step, we may move farther away from the optimum, for example, for a convex function at an optimal point $\boldsymbol{\theta}_*$ obeying $\nabla f(\boldsymbol{\theta}_*) = 0$.

It can be shown that if we choose the step size as $1/k$, then the stochastic gradient method also converges to a minimum of a convex function, under some conditions on the class of functions, however, often at a slower rate, that depends on the class of functions. With a constant step size, the stochastic gradient method converges to a ball around the optimal solution for some convex functions. The larger the step size, the faster the stochastic gradient method converges to this ball. But at the same time, the larger the step size the larger this ball, since the radius of the ball is proportional to the step size. In practice, it is therefore common to choose a decaying step size schedule, which balances those two effects. See the following section 3.5.1 for an analysis for the stochastic gradient with a decaying step size.

Finally, if we minimize a non-differentiable function, we can work with subgradients instead of gradients just like with the subgradient method.

3.5.1 Convergence for a class of functions

We next discuss a standard convergence result of the stochastic gradient method from [Nem+09; WR22; RSS12]. This result requires a second moment bound of the stochastic gradient, specifically we assume that the gradient is (M, B)-bounded in that for all $\boldsymbol{\theta}$,

$$\mathbb{E}\left[\|G(\boldsymbol{\theta})\|_2^2\right] \leq M^2\|\boldsymbol{\theta} - \boldsymbol{\theta}_*\|_F^2 + B^2, \tag{3.10}$$

where $\boldsymbol{\theta}_*$ is the minimizer of L.

Proposition 4 *Let L be m-strongly convex (i.e., $g(\boldsymbol{\theta}) = L(\boldsymbol{\theta}) - \frac{m}{2}\|\boldsymbol{\theta}\|_2^2$ is convex) with minimizer $\boldsymbol{\theta}_*$ and let G be a stochastic gradient that is (M, B)-bounded. Assume we start the SGM iterates at $\boldsymbol{\theta}_0 = 0$, with decaying step size $\eta_k = \frac{2}{m}\frac{1}{2\frac{M^2}{m^2}+k}$. Then*

$$\mathbb{E}\left[\|\boldsymbol{\theta}_k - \boldsymbol{\theta}_*\|_2^2\right] \leq \frac{1}{k-2}\frac{1}{m^2}\left(2M^2\|\boldsymbol{\theta}_0 - \boldsymbol{\theta}_*\|_2^2 + B^2\right).$$

Let us consider again a least-squares example as throughout this chapter:

$$L(\boldsymbol{\theta}) = \frac{1}{N}\|\mathbf{A}\boldsymbol{\theta} - \mathbf{y}\|_2^2 = \frac{1}{N}\sum_{i=1}^{N}(\mathbf{a}^T{}_i\boldsymbol{\theta} - y_i)^2.$$

The stochastic gradient is given by

$$G(\boldsymbol{\theta}) = \nabla(\mathbf{a}_i{}^T\boldsymbol{\theta} - y_i)^2,$$

where i is chosen uniformly at random from all indices. It can be verified that the stochastic gradient is (M, B)-bounded with

$$M = \left(\frac{2}{N}\sum_{i=1}^{N} \|\mathbf{a}_i\|_2^2\right)^{1/2}, \quad B = \left(\frac{2}{N}\sum_{i=1}^{N} \|\mathbf{a}_i\|_2^2 \|\boldsymbol{\theta}_i^* - \boldsymbol{\theta}_*\|_2^2\right)^{1/2},$$

where $\boldsymbol{\theta}_i^*$ is a minimizer of $(\mathbf{a}^T{}_i\boldsymbol{\theta} - y_i)$. Thus, for this least-squares example the number of iterations required by the SGM to obtain an ϵ-accurate solution in expectation is proportional to $1/\epsilon^2$, while the number of iterations required by gradient descent is proportional to $\log(1/\epsilon)$ (see Equation (3.3)). Thus the convergence rate by gradient descent for this example is significantly faster.

As a side note, if the minima of the individual functions, $(\mathbf{a}^T{}_i\boldsymbol{\theta} - y_i)$, coincide with the minimizer of the objective L, then the SGM for this least-squares example (also called randomized Kaczmarz algorithm) achieves a linear rate of convergence, significantly faster than the $1/k$ convergence.

While the stochastic gradient method often converges significantly slower, computing the stochastic gradient $G(\boldsymbol{\theta})$ is by a factor of N cheaper than computing the gradient of the objective L, thus the overall computational cost to reach an ϵ-accurate solution can be faster than for gradient descent.

Proof of Proposition 4: We start by upper bounding the expectation $e_k = \mathbb{E}\left[\|\boldsymbol{\theta}_k - \boldsymbol{\theta}_*\|_2^2\right]$. We write $G_k(\boldsymbol{\theta}_k) = G(\boldsymbol{\theta}_k)$ for the stochastic gradient at iteration k to emphasize that at each iteration, the gradient is drawn independently from the gradient at other iterations. First note that

$$\begin{aligned}\|\boldsymbol{\theta}_{k+1} - \boldsymbol{\theta}_*\|_2^2 &= \|\boldsymbol{\theta}_k - \eta_k G_k(\boldsymbol{\theta}_k) - \boldsymbol{\theta}_*\|_2^2 \\ &= \|\boldsymbol{\theta}_k - \boldsymbol{\theta}_*\|_2^2 - 2\eta_k G_k(\boldsymbol{\theta}_k)^T(\boldsymbol{\theta}_k - \boldsymbol{\theta}_*) + \eta_k^2\|G_k(\boldsymbol{\theta}_k)\|_2^2.\end{aligned}$$

The expectation of the middle term above is

$$\mathbb{E}\left[G_k(\boldsymbol{\theta}_k)^T(\boldsymbol{\theta}_k - \boldsymbol{\theta}_*)\right] = \mathbb{E}\left[\mathbb{E}\left[G_k(\boldsymbol{\theta}_k)|\boldsymbol{\theta}_k\right]^T(\boldsymbol{\theta}_k - \boldsymbol{\theta}_*)\right] = \mathbb{E}\left[\nabla L(\boldsymbol{\theta}_k)^T(\boldsymbol{\theta}_k - \boldsymbol{\theta}_*)\right], \tag{3.11}$$

where we used that the random variable G_k is independent of the random variable $G_i, i < k$, since each stochastic gradient is drawn independently, and is therefore independent of the iterate $\boldsymbol{\theta}_k$. Thus, iterating the expectation allows us to replace the stochastic gradient by the gradient.

Using the assumption that the stochastic gradient G is (M, B)-bounded yields

$$e_{k+1} \leq (1 + \eta_k^2 M^2)e_k - 2\eta_k \mathbb{E}\left[\nabla L(\boldsymbol{\theta}_k)^T(\boldsymbol{\theta}_k - \boldsymbol{\theta}_*)\right] + \eta_k^2 B^2. \tag{3.12}$$

With $\nabla L(\boldsymbol{\theta}_*) = 0$, we have that

$$\nabla L(\boldsymbol{\theta}_k)^T(\boldsymbol{\theta}_k - \boldsymbol{\theta}_*) = \nabla L(\boldsymbol{\theta}_k) - \nabla L(\boldsymbol{\theta}_*)^T(\boldsymbol{\theta}_k - \boldsymbol{\theta}_*) \geq m\|\boldsymbol{\theta}_k - \boldsymbol{\theta}_*\|_2,$$

where the inequality follows from L being m-strongly convex. Application of this inequality to (3.12) yields

$$e_{k+1} \leq (1 + \eta_k^2 M^2 - 2\eta_k m)e_k + \eta_k^2 B^2. \tag{3.13}$$

Since we choose the step size as $\eta_k = \frac{2}{m}\frac{1}{2\beta+k}$ with $\beta = \frac{M^2}{m^2}$, we have $\eta_k \leq \frac{m}{M^2}$. Then Equation (3.13) yields

$$e_{k+1} \leq (1 - \eta_k m)e_k + \eta_k^2 B^2.$$

Unrolling the iterations gives

$$\begin{aligned}
e_k &\leq (1 - \eta_{k-1}m)e_{k-1} + \eta_{k-1}^2 B^2 \\
&\leq (1 - \eta_{k-1}m)(1 - \eta_{k-2}m)e_{k-2} + (1 - \eta_{k-1}m)\eta_{k-2}^2 B^2 + \eta_{k-1}^2 B^2 \\
&\leq \prod_{i=0}^{k-1}(1 - \eta_i m)e_0 + B^2 \sum_{i=0}^{k-1} \eta_i^2 \prod_{j=i+1}^{k-1}(1 - \eta_j m) \\
&= \prod_{i=0}^{k-1}\left(1 - \frac{2}{2\beta + i}\right)e_0 + B^2 \sum_{i=0}^{k-1}\left(\frac{1}{m}\frac{2}{2\beta + i}\right)^2 \prod_{j=i+1}^{k-1}\left(1 - \frac{2}{2\beta + j}\right).
\end{aligned}$$

Using that $\prod_{j=i}^{k}\left(1 - \frac{2}{j}\right)^2 = \frac{(i-2)(i-1)}{(k-1)k}$, we get

$$\begin{aligned}
e_k &\leq \frac{(2\beta - 2)(2\beta - 1)}{(k-2)(k-1)}e_0 + \frac{B^2}{m^2}\sum_{i=0}^{k-1}\left(\frac{2}{2\beta + i}\right)^2 \frac{(2\beta + i - 1)(2\beta + i)}{(k-2)(k-1)} \\
&\leq \frac{4\beta^2}{(k-2)^2}e_0 + \frac{B^2}{m^2}\frac{1}{k-2} \\
&\leq \frac{1}{k-2}\left(2\beta e_0 + \frac{B^2}{m^2}\right),
\end{aligned}$$

which concludes the proof. □

3.6 Momentum and acceleration

We saw that when gradient descent is applied to a quadratic function, then $O(M/m \log(1/\epsilon))$ many iterations are sufficient to obtain an ϵ-accurate solution, where M and m are the largest and smallest singular values of the design matrix. We now briefly discuss a modification of the gradient method that results in a major reduction in iteration complexity, to $O(\sqrt{M/m} \log(1/\epsilon))$ many iterations for an ϵ-accurate solution. This modification/method or class of modifications/methods is called acceleration or momentum. Momentum methods can achieve significant speedups at almost no additional compute costs and are therefore very popular. The core idea of momentum methods is to accelerate gradient methods by considering past gradients in its updates.

For minimizing an objective function $L(\mathbf{x})$, the momentum method first computes the update direction

$$\mathbf{v}_k = \beta \mathbf{v}_{k-1} + (1-\beta)\nabla L(\mathbf{x}_k),$$

and then performs the update

$$\mathbf{x}_{k+1} = \mathbf{x}_k - \eta \mathbf{v}_k.$$

Here, η is the step size parameter and β is the momentum coefficient.

Another instance of a momentum method is the heavy-ball method. The heavy-ball method involves a term that adds a fraction of the previous parameter update, which aligns with the momentum concept of using past gradients for the current update. For minimizing an objective function $L(\mathbf{x})$, it performs the updates

$$\mathbf{x}_{k+1} = \mathbf{x}_k - \eta_k \nabla L(\mathbf{x}_k) + \beta_k(\mathbf{x}_k - \mathbf{x}_{k-1}).$$

Here, $\mathbf{x}_k - \mathbf{x}_{k+1}$ is the momentum term which is simply the previous step of the heavy-ball method. On a quadratic, the heavy-ball method requires $O(\sqrt{M/m} \log(1/\epsilon))$ many iterations to achieve accuracy ϵ; this is by a factor $\sqrt{M/m}$ better than what gradient descent achieves.

A very popular and practically successful method that also attains the rate $O(\sqrt{M/m} \log(1/\epsilon))$ on quadratics is Nesterov's accelerated gradient *method*, which we present next. The accelerated gradient method modifies gradient descent by adding a momentum term:

$$\mathbf{y}_{k+1} = \mathbf{x}_k - \frac{1}{M}\nabla f(\mathbf{x}_k),$$
$$\mathbf{x}_{k+1} = (1-\gamma_k)\mathbf{y}_{k+1} + \gamma_k \mathbf{y}_k$$
$$= \mathbf{y}_{k+1} + \gamma_k(\mathbf{y}_k - \mathbf{y}_{k+1}).$$

So the algorithm performs a gradient descent step to obtain $\mathbf{y}_{k+1}$, but then goes a little further (due to $1-\gamma_k > 1$) and adds the previous direction $\mathbf{y}_k$. In Nesterov's formulation, the rates are defined by

$$\kappa_0 = 0, \quad \kappa_k = \frac{1+\sqrt{1+4\kappa_{k-1}^2}}{2}, \quad \gamma_k = \frac{1-\kappa_k}{\kappa_{k+1}} \leq 0. \tag{3.14}$$

3.7 Adaptive gradient descent

For training deep neural networks, first-order methods that adapt the learning rate are very popular, in particular the adaptive moment estimation algorithm (Adam) [KB15], discussed in the next section. Adam builds on the adaptive gradient method (AdaGrad) [DHS11] discussed next. By adapting to the learning rate, adaptive gradient methods can achieve faster convergence, and are less sensitive to the choice of the step size.

Suppose we want to minimize the objective function $L(\mathbf{x})$. At iteration k, AdaGrad computes a gradient $\mathbf{g}_k$ or a stochastic gradient of the objective function $L(\mathbf{x}_k)$, and computes an update as

$$\mathbf{x}_{k+1} = \mathbf{x}_k - \eta \operatorname{diag}(\mathbf{h}_k)^{-1}\mathbf{g}_k,$$

where η is a step size parameter, and the i-th entry of $\mathbf{h}_k$ is the squared sum of the prior gradients, i.e.,

$$h_{k,i} = \sqrt{\sum_{\ell=1}^{k} g_{\ell,i}^2 + \epsilon}.$$

Here, ϵ is a small numerical constant added for numerical stability.

AdaGrad adjusts the step size for each parameter individually based on its past gradients. If the past gradients for a coordinate are large, then the step size associated with the respective parameter is smaller.

In a neural network, for example, we sometimes have the situation that some layers of the network require a very small step size to adapt, and others a large step size, as the corresponding gradients are very small and large. In such a scenario, gradient descent can fail to learn a well-performing network as the parameters of the network with the small gradients is not changing much during training. AdaGrad can effectively adapt to such a situation, and can also lead to better convergence. We next illustrate this with an example.

3.7.1 Convergence on a diagonal least-squares problem

Let us study the convergence of AdaGrad on a simple optimization problem, to obtain intuition on the behaviour of AdaGrad versus gradient descent. We consider a least-squares problem with diagonal matrix, i.e.,

$$L(\mathbf{x}) = \frac{1}{2}\|\mathbf{\Sigma x} - \mathbf{y}\|_2^2, \tag{3.15}$$

where $\mathbf{\Sigma}$ is a diagonal matrix. For this setup, the AdaGrad updates only have entry-wise dependencies, and as we see next, the convergence does not depend on the singular values of the matrix, as it does for gradient descent.

Proposition 5 *Let $r_{k,i} = \sigma_i x_{k,i} - y_i$ be the residual of the* i*-th coordinate of the k-th iteration of AdaGrad, where σ_i is the i-th entry of the diagonal matrix $\mathbf{\Sigma}$, and suppose that $\mathbf{y} = \mathbf{\Sigma x}^*$. Suppose the step size obeys $\eta \leq |x_{0,i} - x_i^*|$. Then the iterates of AdaGrad converge linearly as*

$$|r_{k,i}| \leq \left(1 - \frac{\eta}{(x_{0,i} - x_i^*)^2}\sqrt{4/3}\right)^{k-1} r_{0,i}.$$

The statement demonstrates that on a quadratic with diagonal entries, AdaGrad adapts well to the different singular values, in that the convergence rate does not depend on the singular value σ_i associated with the i-th coordinate.

In contrast, the convergence rate of gradient descent for this problem depends on the singular value associated with each coordinate. Specifically, the residuals of gradient descent converge as

$$r_{k,i} \leq \left(1 - \eta\sigma_i^2\right)^{k-1} r_{0,i},$$

which can be much slower for this example if some of the singular values are small and others are large.

Proof of Proposition 5: Let us drop the index i for notational convenience. The gradient is given by

$$g_k = \sigma(\sigma x_k - y) = \sigma r_k,$$

where $r_k = \sigma x_k - y$ is the residual at the k-th iteration. We have that

$$\begin{aligned}
r_{k+1} &= \sigma x_{k+1} - y \\
&= \sigma\left(x_k - \frac{\eta}{h_k} g_k\right) - y \\
&= \sigma\left(x_k - \frac{\eta}{h_k} \sigma r_k\right) - y \\
&= \left(1 - \frac{\eta\sigma^2}{h_k}\right) r_k.
\end{aligned}$$

We next provide a coarse bound of the term h_k. Towards this goal, note that

$$\begin{aligned}
h_k &= \sqrt{g_0^2 + \ldots + g_{k-1}^2} \\
&= \sigma\sqrt{r_0^2 + \ldots + r_{k-1}^2} \\
&\leq \sigma|r_0|\sqrt{k},
\end{aligned}$$

where the last inequality follows from the loss being decreasing so that $|r_k| \leq |r_0|$. With this naive bound, we can derive a better bound as follows. Application of this naive bound gives, with $\kappa = \eta\sigma/|r_0|$, for notational simplicity, and using that $\kappa \leq 1$, by the assumption on the step size,

$$\begin{aligned}
r_k &\leq \left(1 - \frac{\kappa}{\sqrt{1}}\right)\left(1 - \frac{\kappa}{\sqrt{2}}\right)\cdots\left(1 - \frac{\kappa}{\sqrt{k-1}}\right) r_0 \\
&= r_0 e^{\log\left(1-\frac{\kappa}{\sqrt{1}}\right)+\log\left(1-\frac{\kappa}{\sqrt{2}}\right)+\ldots} \\
&\leq r_0 e^{-\kappa \sum_{i=1}^{k-1} \frac{1}{\sqrt{i}}} \\
&\leq r_0 e^{-\kappa\sqrt{k-1}},
\end{aligned}$$

where the inequality follows from $\log(z) \leq z - 1$ for all $z > 1$. With this we can get the better bound on h_k as

$$\begin{aligned}
h_k^2 &= \sigma^2\left(r_0^2 + \ldots + r_{k-1}^2\right) \\
&\leq \sigma^2 r_0^2\left(1 + \sum_{\ell=0}^{k-2} e^{-2\kappa\sqrt{\ell}}\right) \\
&\overset{(i)}{\leq} \sigma^2 r_0^2\left(1 + \frac{1}{2\kappa^2}\right) \\
&\overset{(ii)}{\leq} \sigma^2 r_0^2 \frac{3}{4\kappa^2},
\end{aligned}$$

were (ii) follows from the assumption $\kappa \leq 1/2$. Inequality (i) follows from the bound, for $z > 0$,

$$\int_0^k z^{\sqrt{x}} dx = \frac{2}{\ln(z)^2} z^{\sqrt{x}} \left(\ln(z)\sqrt{x} - 1\right) |_{x=0}^k$$
$$= \frac{2}{\ln(z)^2} \left(1 - z^{\sqrt{k}}(1 + \ln(1/z))\right) \leq \frac{2}{\ln(z)^2}.$$

This yields the better bound

$$h_k \leq r_0^2 \sqrt{3/4}.$$

With the better bound we get

$$r_{k+1} \leq \left(1 - \frac{\eta\sigma^2}{h_k}\right) r_k \leq \left(1 - \frac{\eta\sigma^2}{r_0^2}\sqrt{4/3}\right) r_k,$$

and using that $r_0 = \sigma(x_0 - x^*)$ concludes the proof of the proposition. □

3.8 Adam

The adaptive moment estimation algorithm (Adam) [KB15] for optimizing a loss L $(\mathbf{x})$ combines the per-parameter update adjustment of AdaGrad with momentum. Adam maintains an exponentially decaying average of past gradients as a momentum term. Adam also maintains exponentially decaying average of squared gradients as a second moment estimate similar to AdaGrad. The update rules are as follows:

Compute a first moment estimate as

$$\hat{\mathbf{m}}_t = \frac{1}{\beta_1^t} \mathbf{m}_t, \quad \mathbf{m}_t = \beta_1 \mathbf{m}_{t-1} + (1 - \beta_1)\mathbf{g}_t \tag{3.16}$$

and a second moment estimate as

$$\hat{\mathbf{v}}_t = \frac{1}{\beta_2^t} \mathbf{v}_t, \quad \mathbf{v}_t = \beta_2 \mathbf{v}_{t-1} + (1 - \beta_2)\mathbf{g}_t^2. \tag{3.17}$$

The parameters are then updated as

$$\mathbf{x}_{t+1} = \mathbf{x}_t - \eta \frac{\hat{\mathbf{m}}_t}{\sqrt{\hat{\mathbf{v}}_t} + \epsilon}, \tag{3.18}$$

where the division and square-root are entry-wise operations.

The second moment estimate is related to but different from the second moment estimate by AdaGrad; the AdaGrad update rule is to simply accumulate the second moments, i.e., $\mathbf{v}_{t+1} = \mathbf{v}_t + \mathbf{g}_t^2$, while the second moment estimate of Adam is exponentially decaying. In addition, Adam includes a bias correction for the first and second moments.

The default hyperparameters are $\beta_1 = 0.9, \beta_2 = 0.999$, and $\epsilon = 1e - 8$. Adam performs very well for optimizing neural networks, and a large fraction of the current neural networks for image reconstruction are optimized with the Adam algorithm.

3.9 Chapter notes

This chapter only scratches the surface of convex optimization and first-order optimization methods for solving optimization problems.

A standard reference for the fundamentals of convex optimization is the book [BV04]. The book [WR22] covers the fundamentals of optimization algorithms, including gradient descent, momentum, the stochastic gradient method, and associated convergence rates.

Most (convergence) analysis of first-order methods is for convex functions, and it is difficult to analyse the behaviour for non-convex functions, including loss functions involving neural networks. However, in some regimes, for example sufficiently wide neural networks, it is possible to approximate the training of a neural network well with an associated linear function, and analyse the linear function to obtain convergence rates and to understand the behavior of first order methods applied to non-convex functions [JGH18].

4
Sparse modelling and sparse signal reconstruction

Signal models play a pivotal role for solving inverse problems, since reconstruction methods either implicitly or explicitly use signal models. Assume a signal that we are interested in reconstructing lies in a class of signals described by a signal model. Our goal is to reconstruct the signal from a measurement using the prior knowledge of the signal being described by the signal model, and we wish to do so with an algorithm that is sample efficient (i.e., only requires a minimal number of measurements) and is computationally efficient.

An example of such an algorithm is the following. If a n-dimensional signal lies in a d-dimensional subspace, then the n-dimensional signal can be perfectly reconstructed from d many linear measurements, even if the ambient dimension, n, is much larger than the dimension of the subspace, d. Moreover, reconstructing a signal in a subspace from linear measurements consists of solving a linear system of equations which is sample and computationally efficient.

However, while for some signals, a subspace is a good model (for example, the set of images of a given face under varying illumination conditions approximately lies in a 9-dimensional linear subspace [BJ03]), most signals of interest are not described well by a subspace. For example, for natural images, a single subspace is not a good model.

A natural generalization of assuming a signal lies in a subspace is to assume the signal lies in a union of subspaces. Sparse signals, which we discuss in this chapter, are signals that lie in a union of subspaces. Sparse signals can describe natural images well, as discussed in this chapter.

In this chapter, we discuss sparse signals because they can describe many signals of practical interest well, in particular images, and can be recovered sample and computationally efficient.

4.1 Sparse signal models

We say that a signal $\mathbf{x} \in \mathbb{R}^n$ (or $\mathbb{C}^n$) is sparse if the number of non-zero elements, denoted by

$$\|\mathbf{x}\|_0 = |\{x_i : x_i \neq 0\}|$$

Deep Learning for Computational Imaging. Reinhard Heckel, Oxford University Press. © Reinhard Heckel (2025).
DOI: 10.1093/oso/9780198947172.003.0004

is small relative to the ambient dimension n. The meaning of small in this context is application-dependent. Often small means that the number of non-zeros of $\mathbf{x}$ is a small fraction of the ambient dimension, or even that the number of non-zeros is sub-linear in the ambient dimensions, for example, that it scales like $\sqrt{n}$ or $\log(n)$.

We say that a signal $\mathbf{x}$ is s-sparse if it has s many non-zero coefficients, i.e., $\|\mathbf{x}\|_0 = s$. In practice, signals are typically not *exactly* sparse, but can be approximated well with a sparse signal. A vector is approximated well with a sparse signal if its best s-sparse approximation error, defined as

$$\sigma_s(\mathbf{x}) = \inf\{\|\mathbf{x} - \mathbf{x}_s\|_2 : \mathbf{x}_s \text{ is } s - \text{sparse}\},$$

decays quickly in s. While we focus on the reconstruction of exactly sparse signals, the algorithms we consider also work well for estimating approximately sparse signals. In addition, many reconstruction guarantees extend from exactly sparse signals to compressible signals.

Often a signal of interest is not sparse itself, but is sparse when represented as a linear combination of a set of vectors

$$\mathcal{D} = \{\mathbf{d}_1, \ldots, \mathbf{d}_N\} \subset \mathbb{R}^n$$

called a dictionary. Specifically, we say that the signal $\tilde{\mathbf{x}} \in \mathbb{R}^n$ is s-sparse in the dictionary $\mathcal{D}$ if it has a representation

$$\tilde{\mathbf{x}} = \mathbf{d}_1 x_1 + \ldots + \mathbf{d}_N x_N,$$

with only s coefficients $x_1, \ldots, x_N$ being non-zero. For example, images are known to have an (approximately) sparse representation in the *discrete Cosine basis* (as demonstrated below) and typically an even sparser representation in a wavelet dictionary. As a second example, a Radar signal received by a Radar receiver is sparse in the dictionary consisting of time and frequency shifted versions of the transmitted signal.

A dictionary that is able to represent a given class of signal (natural images, music,. . .), is typically built based on domain knowledge and expert intuition. It is also possible to learn a suitable dictionary for a given dataset, which is known as dictionary learning [AEB06].

A dictionary can be an *orthonormal basis* or an overcomplete dictionary, we discuss both next.

Orthonormal basis

A set of vectors $\mathcal{D}$ is called an orthonormal basis for the Euclidean space $\mathbb{R}^n$, if $\mathcal{D}$ is a basis for $\mathbb{R}^n$, i.e., the vectors in $\mathcal{D}$ are linearly independent and each vector

(a) Original image (b) 10% coeffs (c) 1% coeffs

Figure 4.1 (a) An original, uncompressed image. (b) and (c): We take the image, represent it in terms of its DCT coefficients, and set the smallest 90% and 99% of the coefficients to zero. Even when only retaining the 10% largest coefficients, the image is close to the original image. This is the case since most energy of the signal is represented in few coefficients, which can be seen when inspecting the distribution of the DCT coefficients.

in $\mathbb{R}^n$ can be represented as a linear combination of vectors in $\mathcal{D}$, and all vectors are normalized and orthogonal to each other, i.e., $\|\mathbf{x}_i\|_2 = 1$, for all $\mathbf{x}_i \in \mathcal{D}$ and $\langle \mathbf{x}_i, \mathbf{x}_j \rangle = 0$ for all pairs $\mathbf{x}_i, \mathbf{x}_j \in \mathcal{D}$ with $i \neq j$.

An important example of an orthonormal basis for the complex Euclidean space $\mathbb{C}^n$ is the discrete Fourier transform (DFT). The DFT maps vectors in $\mathbb{C}^n$ to their decomposition in terms of discrete sinusoids:

$$\mathcal{D} = \{\mathbf{f}_0, \ldots, \mathbf{f}_{n-1}\}, \quad \text{where} \quad \mathbf{f}_k = \frac{1}{\sqrt{n}} \begin{bmatrix} 1 \\ e^{i2\pi k/n} \\ e^{i2\pi 2k/n} \\ \ldots \\ e^{i2\pi(n-1)k/n} \end{bmatrix}. \tag{4.1}$$

Images are (approximately) sparse in a closely related real valued dictionary, the discrete Cosine transform (DCT). The DCT decomposes a signal into a sum of cosines of different frequencies. To see that natural images are approximately sparse in the DCT, consider the image shown on the left of Figure 4.1. Its DCT coefficients ordered by magnitude decay rather quickly, as depicted in Figure 4.1. We can therefore compress an image by only retaining a fraction of the largest DCT coefficients. The JPEG format and other lossy compression schemes are based on this idea. See Figure 4.1 for the image obtained from a fraction of the most significant DCT coefficients.

Overcomplete dictionaries.

A dictionary that spans $\mathbb{R}^n$ is said to be overcomplete if removing a vector (or several vectors) from the dictionary results in a dictionary that is still complete in that it spans $\mathbb{R}^n$. Overcomplete dictionaries can be more efficient than bases in sparsely representing a class of signals. To see this, let us consider an example.

Consider the concatenation of the Fourier basis and the standard basis

$$\mathcal{D} = \{\mathbf{f}_0, \ldots, \mathbf{f}_{n-1}, \mathbf{e}_0, \ldots, \mathbf{e}_{n-1}\}.$$

Here, $\mathbf{f}_k$ is a column of the discrete Fourier transform defined in Equation (4.1), and $\mathbf{e}_k$ is the unit vector with k-th entry equal to 1. Consider the signal $\mathbf{f}_0 + \mathbf{e}_0$. In both the Fourier and the standard basis, this signal has a maximally dense representation, meaning that n coefficients are non-zero. However in the dictionary $\mathcal{D}$, it has a 2-sparse representation. This illustrates that it is easier to find an overcomplete dictionary than it is to find an orthonormal basis in which a class of signals is sparse.

Suppose we are given a signal $\tilde{\mathbf{x}}$ and a dictionary and we want to find a representation $\tilde{\mathbf{x}} = \mathbf{D}\mathbf{x}$, where $\mathbf{D}$ is a matrix with the vectors of the dictionary as columns, and $\mathbf{x}$ is sparse. Since the dictionary is overcomplete, there are many solutions $\mathbf{x}$ that obey $\tilde{\mathbf{x}} = \mathbf{D}\mathbf{x}$, finding a sparse solution is a sparse signal recovery problem.

4.2 Problem formulation and applications

We consider the problem of reconstructing an s-sparse signal $\mathbf{x} \in \mathbb{R}^n$ from m linear measurements

$$\mathbf{y} = \mathbf{A}\mathbf{x} \in \mathbb{R}^m, \tag{4.2}$$

where $\mathbf{A} \in \mathbb{R}^{m \times n}$ is a given measurement matrix. We focus on problem setups where the measurement matrix has fewer rows than columns, and therefore we cannot simply invert the matrix to reconstruct the unknown signal.

There are also interesting problem setups where the measurement matrix does not have fewer rows than columns but is ill-conditioned (i.e., the fraction of the largest over smallest singular value is large) and thus it is not easy to stably invert the matrix.

Sparse recovery has many applications in imaging and beyond imaging. Here are a few examples:

- **Denoising.** Suppose we observe the noisy version of a signal $\mathbf{x}$, i.e., $\mathbf{y} = \mathbf{x} + \mathbf{e}$, where $\mathbf{e}$ is noise. If $\mathbf{x}$ is sparse in some dictionary in which the noise is *not* sparse (e.g., if the noise is Gaussian this is true with high probability, regardless of the dictionary), then a large part of the noise can be removed by finding a sparse approximation of $\mathbf{x}$. An image can be denoised using this approach; Elad and Aharon [EA06] proposed such a denoising approach based on a learned dictionary.

- **Data separation**: Suppose we are given a vector that we know has multiple components. For example, suppose that we observe a signal $\mathbf{y} = \mathbf{x}_1 + \mathbf{x}_2$, and we know that the signal $\mathbf{x}_1$ is sparse in the discrete cosine transform and $\mathbf{x}_2$ is sparse in the standard basis. Then we can find a sparse approximation of the signal $\mathbf{y}$ in terms of cosines and spikes, and this enables us to separate the two signals, i.e., reconstruct $\mathbf{x}_1$ and $\mathbf{x}_2$ perfectly. In audio signal processing this allows to separate pure harmonics from spikes, and in image processing, this enables us to remove scratches from natural images.
- **Compressed sensing.** We are often interested in reconstructing an object from few measurements or samples only, as taking a measurement has a cost associated with it. Suppose we take linear measurements of a signal $\tilde{\mathbf{x}}$ that is sparse in the basis $\mathbf{D} \in \mathbb{R}^{n \times n}$ with a measurement matrix $\tilde{\mathbf{A}} \in \mathbb{R}^{m \times n}$, i.e.,

$$\mathbf{y} = \tilde{\mathbf{A}}\tilde{\mathbf{x}} = \underbrace{\tilde{\mathbf{A}}\mathbf{D}}_{\mathbf{A}}\mathbf{x},$$

  and suppose the number of measurements, m, is much smaller than the ambient dimension of the signal, n.

  Reconstruction of the signal $\mathbf{x}$ without making assumptions on the signal is impossible since the matrix $\mathbf{A}$ is not invertible. However, if the signal $\mathbf{x}$ is sparse, we can reconstruct it, and with the reconstructed signal we can reconstruct the signal of interest according to $\tilde{\mathbf{x}} = \mathbf{D}\mathbf{x}$.

  This idea has applications in magnetic resonance imaging (MRI), a medical imaging technique, discussed in the Introduction (Section 1.1.2). The goal in MRI is to obtain an accurate image from few measurements, since collecting measurements is time-consuming, costly, and a person in an MRI scanner cannot remain motionless for long. An MRI scanner collects measurements of an image $\tilde{\mathbf{x}}$ in the Fourier domain according to $\mathbf{y} = \mathbf{MF}\tilde{\mathbf{x}}$, where $\mathbf{F}$ is the Fourier transform and $\mathbf{M} \in \mathbb{R}^{m \times n}$ is a mask that selects a few rows of the Fourier transform of the image, $\mathbf{F}\tilde{\mathbf{x}}$. Now assume that the image is sparse in the wavelet domain, i.e., $\tilde{\mathbf{x}} = \mathbf{W}\mathbf{x}$, with $\mathbf{x}$ sparse and $\mathbf{W}$ being a Wavelet dictionary. Thus, the measurement is $\mathbf{y} = \mathbf{MFW}\mathbf{x}$, and using the assumption that $\mathbf{x}$ is sparse, we can reconstruct the sparse representation $\mathbf{x}$ from the measurement $\mathbf{y}$. From the sparse representation we can compute the image as $\tilde{\mathbf{x}} = \mathbf{W}\mathbf{x}$.
- **Statistics and machine learning.** The goal of regression is to predict an outcome based on input data. A common model is

$$\mathbf{y} = \mathbf{A}\mathbf{x} + \mathbf{e},$$

  where $\mathbf{y}$ is the outcome, $\mathbf{A}$ is called the design or predictor matrix, $\mathbf{x}$ is the model parameter, and $\mathbf{e}$ is a random noise vector. For example, $\mathbf{y}$ could be the probability of a patient having a disease and the columns of $\mathbf{A}$ could be data vectors corresponding to blood pressure, weight, etc. Often, only a small number of the parameters (columns of $\mathbf{A}$) contribute to the effect. Therefore, fitting a model by determining a parameter vector $\mathbf{x}$ that describes the outcome $\mathbf{y}$ well is a sparse approximation problem.

4.3 Algorithms for sparse reconstruction

In this section we discuss algorithms for the reconstruction of a sparse signal $\mathbf{x} \in \mathbb{R}^n$ from a noiseless measurement

$$\mathbf{y} = \mathbf{A}\mathbf{x}.$$

Given that we know that the vector $\mathbf{x}$ is sparse, an intuitive reconstruction approach is to find the vector with the smallest number of non-zeros that is consistent with the measurement:

$$\hat{\mathbf{x}} = \arg\min_{\mathbf{x}} \|\mathbf{x}\|_0 \text{ subject to } \mathbf{y} = \mathbf{A}\mathbf{x}.$$

This optimization problem is called ℓ_0-norm minimization (even though $\|\cdot\|_0$ is not a norm).

Provided that every set of $2s$ many column of $\mathbf{A}$ is linearly independent, ℓ_0-minimization recovers any s-sparse vector from the observation $\mathbf{y} = \mathbf{A}\mathbf{x}$. This condition is also necessary for any algorithm to reconstruct any s-sparse vector.

If the measurement matrix $\mathbf{A}$ is a Gaussian random matrix, i.e., a matrix with iid Gaussian entries, then this implies that $m = 2s$ measurements are sufficient for ℓ_0-minimization to provably succeed. That follows since every set of $2s$ columns of a wide Gaussian matrix with $m = 2s$ rows is linearly independent with probability one. Note that since s many measurements are necessary to reconstruct $\mathbf{x}$ from $\mathbf{y}$ even when we know the location of the non-zero entries of the vector $\mathbf{x}$, ℓ_0-norm minimization is sample efficient for this scenario.

To see that ℓ_0-norm minimization can reconstruct every s-sparse vector provided that every set of $2s$ column of $\mathbf{A}$ is linearly independent, suppose that $\hat{\mathbf{x}} \neq \mathbf{x}$ is the solution to ℓ_0-minimization with $\|\hat{\mathbf{x}}\|_0 \leq s$. Then

$$\mathbf{A}(\hat{\mathbf{x}} - \mathbf{x}) = \mathbf{0}.$$

Since the difference $\hat{\mathbf{x}} - \mathbf{x}$ has at most $2s$ non-zeros, this implies that there is a set of $2s$ column of $\mathbf{A}$ that is linearly dependent, which contradicts the assumption that every set of $2s$ column of $\mathbf{A}$ is linearly independent.

Unfortunately, the ℓ_0-norm minimization problem is computationally intractable. Specifically, it can be shown to be NP hard, by reducing it to the exact cover by 3-sets problem, which is known to be NP hard. Note that this does *not* say that for a specific $\mathbf{A}$ and $\mathbf{y}$ a solution cannot be found efficiently; the statement is that that an algorithm that solves ℓ_0 for any matrix $\mathbf{A}$ and any $\mathbf{y}$ must be intractable.

We next discuss two popular trackable methods to recover a sparse vector: ℓ_1-norm minimization-based approaches and greedy methods.

4.3.1 ℓ_1-norm minimization

Since ℓ_0-norm minimization is computationally infeasible, a natural strategy is to replace the ℓ_0-norm objective with the closest convex norm, which is the ℓ_1 norm:

$$\underset{\mathbf{x}}{\text{minimize}} \|\mathbf{x}\|_1 \text{ subject to } \mathbf{y} = \mathbf{A}\mathbf{x}. \tag{4.3}$$

Since the ℓ_1-norm objective as well as the constraint $\mathbf{y} = \mathbf{A}\mathbf{x}$ are convex, this is a convex optimization problem that can be solved very efficiently with iterative methods, such as projected gradient descent discussed in Chapter 3. Also, see Chapter 3 for the definition of a convex optimization problem; for now the important property of a convex optimization problem is that it can be solved relatively efficiently.

We call the optimization problem in Equation (4.3) ℓ_1-norm minimization or basis pursuit (BP) in the following. The name basis pursuit originates from the optimization problem it describes, i.e., to find the best basis to represent a signal, and appears in early papers on sparse reconstruction, such as the paper by Chen et al. [CDS+98]. Here, basis refers to the vectors (columns of the matrix $\mathbf{A}$) that combined linearly to represent a signal $\mathbf{y}$ as $\mathbf{y} = \mathbf{A}\mathbf{x}$. The process of pursuing the sparse representation that uses the minimal required elements from the basis coined the name basis pursuit.

In the following, we discuss mathematical conditions under which sparse reconstruction with ℓ_1-minimization succeeds.

Figure 4.2 illustrates a situation where ℓ_1-minimization succeeds in recovering a sparse vector, and a situation where it fails. The figure also provides intuition why ℓ_2-minimization often yields non-sparse solutions: The ℓ_1-ball is more concentrated around the axis than the ℓ_2-ball, therefore it is more likely that minimizing the ℓ_1-norm yields a sparse solution.

We next formalize the intuition from Figure 4.2, to obtain an recovery condition, and to understand under which conditions ℓ_1-minimization recovers a sparse vector. ℓ_1-minimization recovers $\mathbf{x}$ from $\mathbf{y} = \mathbf{A}\mathbf{x}$, if all vectors $\tilde{\mathbf{x}}$ in the set

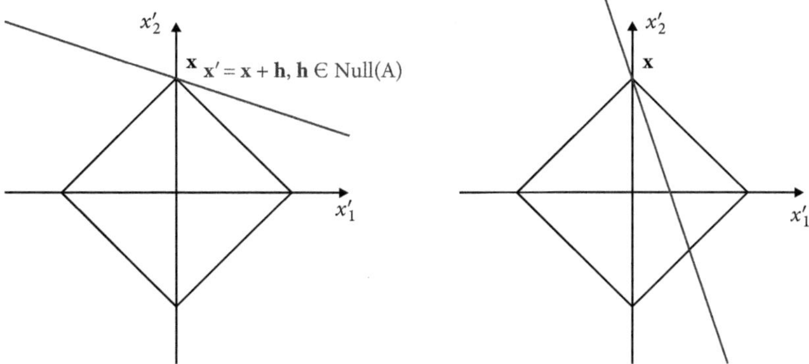

Figure 4.2 The ℓ_1-ball together with the subspace of feasible solutions $\{\mathbf{x}' : \mathbf{A}\mathbf{x}' = \mathbf{y}\} = \{\mathbf{x} + \mathbf{h} : \mathbf{h} \in \text{Null}(\mathbf{A})\}$. If the tangent cone at $\mathbf{x}$ of the ℓ_1-norm and the null space of $\mathbf{A}$ only intersect in $\{\mathbf{0}\}$, then ℓ_1-minimization recovers $\mathbf{x}$. Left: ℓ_1-minimization recovers $\mathbf{x}$; Right: ℓ_1-minimization fails to recover $\mathbf{x}$.

$\{\tilde{\mathbf{x}} \in \mathbb{R}^n \setminus \{\mathbf{x}\} : \mathbf{y} = \mathbf{A}\tilde{\mathbf{x}}\}$ have a larger ℓ_1-norm than $\mathbf{x}$, i.e.,

$$\|\tilde{\mathbf{x}}\|_1 > \|\mathbf{x}\|_1.$$

Writing $\mathbf{h} = \tilde{\mathbf{x}} - \mathbf{x}$, this is equivalent to

$$\|\mathbf{x} + \mathbf{h}\|_1 > \|\mathbf{x}\|_1 \text{ for all } \mathbf{h} \in \text{Null}(\mathbf{A}), \tag{4.4}$$

where $\text{Null}(\mathbf{A}) = \{\mathbf{x} \in \mathbb{R}^n : \mathbf{A}\mathbf{x} = 0\}$ is the nullspace of the matrix $\mathbf{A}$. Thus, ℓ_1-minimization succeeds if the intersection of the tangent cone of the ℓ_1-norm at $\mathbf{x}$, defined as

$$\mathcal{T}(\mathbf{x}) := \{\mathbf{z} \in \mathbb{R} : \|\mathbf{x} + t\mathbf{z}\|_1 \leq \|\mathbf{x}\|_1, \text{for some } t > 0\},$$

only intersects with the null space of $\mathbf{A}$ at $\mathbf{0}$, i.e.,

$$\mathcal{T}(\mathbf{x}) \cap \text{Null}(\mathbf{A}) = \{\mathbf{0}\}. \tag{4.5}$$

This condition is also necessary for ℓ_1-minimization to succeed.

Typically, the Condition (4.5) is difficult to check, and it also depends on $\mathbf{x}$, the signal to be reconstructed. However, Condition (4.5) can be shown to hold true with high probability for Gaussian random matrices for any fixed sparse signal $\mathbf{x}$. The corresponding proof of this result is based on Gordon's lemma which characterizes the likelihood that a random subspace misses a cone [Cha+12]. This proof strategy allows to obtain a very accurate recovery condition, namely that with high probability

$$m > 2s\log(n/s) \tag{4.6}$$

measurements are sufficient for ℓ_1-minimization to recover an s-sparse $\mathbf{x}$ from a measurement $\mathbf{y}$ (this holds in the large n and mildly large s regime, see [FR13, Thm. 9.18] for a precise statement). This is a so-called non-uniform recovery result, i.e., the statement guarantees that a fixed, sparse $\mathbf{x}$ can be recovered with high probability. A uniform recovery result, discussed later, is stronger in that it guarantees that with high probability, all s-sparse vectors (for s sufficiently small) can be recovered.

4.3.2 Greedy methods

We next introduce the orthogonal matching pursuit (OMP) algorithm [TG07], as an alternative algorithm to ℓ_1-minimization introduced in the previous section. The method is illustrative for a class of so-called greedy methods that are focused on building the support set of the sparse vector iteratively.

OMP algorithm: Input: $\mathbf{y}, \mathbf{A}$, and some sparsity level s. At initialization, set $\widehat{\mathcal{S}} = \{\}$ and define the residual $\mathbf{r}_0 = \mathbf{y}$. For $k = 1, \ldots, s$.

1. Find the column of the matrix $\mathbf{A} = [\mathbf{a}_1, \ldots, \mathbf{a}_n]$ that is most correlated with the residual:

$$i = \arg\max_i |\langle \mathbf{r}_{k-1}, \mathbf{a}_i \rangle|.$$

2. Update the estimated support set: $\widehat{\mathcal{S}} \leftarrow \widehat{\mathcal{S}} \cup \{i\}$.
3. Find the best approximation of $\mathbf{y}$ in terms of columns in $\widehat{\mathcal{S}}$: $\mathbf{x}_{\widehat{\mathcal{S}}} = \arg\min \|\mathbf{y} - \mathbf{A}_{\widehat{\mathcal{S}}}\mathbf{x}_{\widehat{\mathcal{S}}}\|_2$, where $\mathbf{A}_{\widehat{\mathcal{S}}}$ is the sub-matrix of $\mathbf{A}$ consisting of columns indexed by the set $\widehat{\mathcal{S}}$.
4. Update the residual: $\mathbf{r}_k = \mathbf{y} - \mathbf{A}_{\widehat{\mathcal{S}}}\mathbf{x}_{\widehat{\mathcal{S}}}$.

The OMP algorithm is appealing as it is easy to implement and computationally efficient. A conceptual disadvantage of this particular greedy algorithm is that if a "wrong" index gets selected, it stays in the support set. There are other greedy algorithms with similar computational complexity that both exclude and include indices in each step, such as the CoSaMP algorithm [NT09], not discussed here.

4.4 Reconstruction guarantees based on incoherence

Both OMP and ℓ_1-norm minimization work well if the columns of the matrix **A** are sufficiently uncorrelated. In this section, we discuss reconstruction guarantees for OMP and ℓ_1-minimization that are based on the correlation of pairs of columns of the matrix **A**. Those results depend on the largest correlation between pairs of columns of the matrix **A**, defined as the incoherence parameter below.

Definition 4 *A matrix* **A** *with unit norm columns (i.e.,* $\|\mathbf{a}_i\|_2 = 1$*, for all columns* $\mathbf{a}_i$ *of* **A***) is* μ*-incoherent if for all pairs of columns* $\mathbf{a}_i, \mathbf{a}_j$*,* $i \neq j$ *of* **A***,*

$$|\langle \mathbf{a}_i, \mathbf{a}_j \rangle| \leq \mu.$$

A small incoherence parameter guarantees that both OMP and ℓ_1-minimization provably reconstruct a sparse vector:

Theorem 1 (e.g., [Tro04, Corollary 3.6]). *Suppose* **A** *is* μ*-incoherent. Let* **x** *be* s*-sparse with*

$$s < \frac{1}{2\mu}. \tag{4.7}$$

Then the solution to both OMP and ℓ_1*-minimization applied to* $\mathbf{y} = \mathbf{A}\mathbf{x}$ *is* **x**.

Below, we prove Theorem 1 for OMP. An advantage of the coherence-based guarantees is that the coherence of a matrix can easily be computed, and that there are deterministic matrices which minimize the coherence for a given shape of the matrix, and thus maximize the sparsity in Condition (4.7). Moreover, the result guarantees recovery of all s-sparse vectors simultaneously, and is therefore uniform.

As formalized by Theorem 1, the smaller the incoherence parameter, the better for sparse reconstruction. It is not difficult to find a matrix with small incoherence parameter:

Proposition 6 *Consider the matrix* $\mathbf{A} \in \mathbb{R}^{m \times n}$ *constructed by choosing each entry as either* $-1/\sqrt{m}$ *or* $+1/\sqrt{m}$ *with equal probability. This matrix has unit norm columns by construction, and the matrix is incoherent with* $\mu = O(\sqrt{\log(n/\delta)/m})$ *with probability at least* $1 - \delta$*.*

Proof of Proposition 6: For independent random variables Z_ℓ that are ± 1 with equal probability, Hoeffding's inequality states that, for all $\beta > 0$,

$$\mathrm{P}\left[\left|\sum_{\ell=1}^{m}(Z_\ell - \mathbb{E}[Z_\ell])\right| \geq \beta\right] \leq 2e^{-\frac{\beta^2}{2}m}.$$

Next consider the inner product $\langle \mathbf{a}_i, \mathbf{a}_j \rangle = \frac{1}{m}\sum_{\ell=1}^{m} a_{i\ell}a_{j\ell} = \frac{1}{m}\sum_{\ell=1}^{m} Z_\ell$, where Z_ℓ is ± 1 with equal probability, and Hoeffding's inequality applies. Thus we obtain

$$\mathrm{P}\left[\max_{i,j} |\langle \mathbf{a}_i, \mathbf{a}_j \rangle| \geq \mu\right] \leq \sum_{i,j} \mathrm{P}\left[|\langle \mathbf{a}_i, \mathbf{a}_j \rangle| \geq \mu\right] \leq n^2 e^{-\frac{\mu^2}{2}m}.$$

Stated differently, we have that $\max_{i,j} |\langle \mathbf{a}_i, \mathbf{a}_j \rangle| \leq \mu = O\left(\sqrt{\log(n/\delta)/m}\right)$ with probability at least $1 - \delta$, as claimed. □

There is however a limit on how incoherent a matrix can be: The coherence of a matrix $\mathbf{A} \in \mathbb{R}^{m \times n}$ obeys the inequality [FR13, Thm. 5.7]:

$$\mu \geq \sqrt{\frac{n-m}{m(n-1)}}.$$

If the number of columns n is a multiple of the number of rows m, the lower bound becomes $\mu \geq c/\sqrt{m}$. There are deterministic constructions that meet this lower bound with equality, thus the bound is tight.

This lower bound on the incoherence implies that the reconstruction guarantee based on incoherence is very pessimistic, and requires the number of measurements to be quadratic in the sparsity (instead of near-linear as Condition (4.6) requires). Since the coherence is lower bounded by $c/\sqrt{m}$ (for, say $n \geq 2m$), Condition (4.7) can only hold when

$$s \leq \frac{\sqrt{m}}{2c}.$$

This is sometimes called the square-root bottleneck.

We hasten to add that the coherence condition is by no means necessary; both ℓ_1-minimization and *adaptions* of OMP provably reconstruct an s-sparse signal from $O(s\log(n/s))$ measurements under certain conditions (e.g., when $\mathbf{A}$ is a Gaussian random matrix), as we see in the coming sections.

4.4.1 Proof of Theorem 1 for OMP

Let $\mathcal{S}$ be the support of $\mathbf{x}$, i.e., the set of indices of the non-zero coefficients of $\mathbf{x}$. We prove the result by induction. Suppose that at iteration $k-1$ the set selected by the OMP algorithm is a subset of $\mathcal{S}$. We show that at k, OMP selects an index in $\mathcal{S}$. To this end, first note that if $\widehat{\mathcal{S}} \subset \mathcal{S}$, then

$$\mathbf{r}_{k-1} = \sum_{i \in \mathcal{S}'} c_i \mathbf{a}_i,$$

where $\mathcal{S}' \subset \mathcal{S}$ and c_i are some coefficients. To see this, note that the residual is

$$\mathbf{r}_{k-1} = \mathbf{A}_{\mathcal{S}} \mathbf{x}_{\mathcal{S}} - \mathbf{A}_{\widehat{\mathcal{S}}} \mathbf{x}_{\widehat{\mathcal{S}}}.$$

Let c_j be the largest coefficient in magnitude. OMP will pick the index j, or another index in $\mathcal{S}'$, if

$$\left|\langle \mathbf{r}_{k-1}, \mathbf{a}_j \rangle\right| > \max_{\ell \notin \mathcal{S}'} \left|\langle \mathbf{r}_{k-1}, \mathbf{a}_\ell \rangle\right|. \tag{4.8}$$

The left hand side (LHS) of (4.8) can be lower bounded as

$$\begin{aligned} \left|\langle \mathbf{r}_{k-1}, \mathbf{a}_j \rangle\right| = \left|\left\langle \sum_{i \in \mathcal{S}'} c_i \mathbf{a}_i, \mathbf{a}_j \right\rangle\right| &\geq |c_j| - \sum_{i \in \mathcal{S}' \setminus \{j\}} |c_i| \left|\langle \mathbf{a}_i, \mathbf{a}_j \rangle\right| \\ &\geq |c_j|(1 - (s-1)\mu) \geq |c_j|(1/2 + \mu), \end{aligned}$$

where the last inequality follows from the assumption $s < \frac{1}{2\mu}$.

The right hand side (RHS) of (4.8) can be upper bounded as:

$$\left|\langle \mathbf{r}_{k-1}, \mathbf{a}_\ell \rangle\right| = \left|\left\langle \sum_{i \in \mathcal{S}'} c_i \mathbf{a}_i, \mathbf{a}_\ell \right\rangle\right| \leq |c_j||\mathcal{S}'|\mu \leq |c_j| s\mu < |c_j| \frac{1}{2}. \tag{4.9}$$

Since $|c_j|\frac{1}{2} < |c_j|(1/2 + \mu)$, it follows that (4.8) holds. This concludes the proof.

4.5 Reconstruction guarantees for individual vectors with ℓ_1-minimization

In this section, we discuss recovery guarantees for ℓ_1-minimization. We consider certain random matrices $\mathbf{A} \in \mathbb{R}^{m \times n}$ and show that, with high probability, ℓ_1-minimization recovers any *fixed* s-sparse vector $\mathbf{x}$ from the measurement

$$\mathbf{y} = \mathbf{A}\mathbf{x}.$$

Results of this kind are often called non-uniform as they only guarantee that a *fixed* sparse vector can be reconstructed with high probability. In contrast, uniform recovery results, discussed later, guarantee that with high probability, every s-sparse vector can be reconstructed with the same realization of a random matrix $\mathbf{A}$.

We start with non-uniform results, as they are easier to prove, and sometimes guarantee reconstruction under milder conditions. We first state a classical result about sub-sampled Fourier matrices by [CRT06a].

Theorem 2 *Suppose the rows of $\mathbf{A} \in \mathbb{C}^{m\times n}$ are chosen uniformly at random from the rows of the DFT matrix $\mathbf{F} \in \mathbb{C}^{n\times n}$, and let $\mathbf{x} \in \mathbb{C}^n$ be s-sparse. If*

$$m \geq c_M s \log n,$$

then, with probability at least $1 - n^{-M}$, the unique solution to ℓ_1-minimization in (4.3) is $\mathbf{x}$. Here, c_M is a scalar that increases in the constant M.

This result is remarkable as it guarantees that an s-sparse signal can be recovered efficiently from a near-linear number of Fourier samples.

This result is possible because the signal, i.e., $\mathbf{x}$ is sparse in the identity basis and we measure that signal by taking inner product with vectors that are incoherent to the identity basis. Specifically, the rows of the Fourier basis are incoherent to the standard basis in the following sense. We say that two orthonormal bases $\{\mathbf{u}_1, \dots, \mathbf{u}_n\}$ and $\{\mathbf{v}_1, \dots, \mathbf{v}_n\}$ are ν-incoherent if

$$\max_{i,j} |\langle \mathbf{u}_i, \mathbf{v}_j \rangle|^2 \leq \frac{\nu}{n}.$$

The Fourier basis $\{\mathbf{f}_1, \dots, \mathbf{f}_n\}$ is maximally incoherent to the standard basis $\{\mathbf{e}_1, \dots, \mathbf{e}_n\}$, i.e.,

$$\max_{i,j} |\langle \mathbf{e}_i, \mathbf{f}_j \rangle|^2 = \frac{1}{n},$$

while trivially the standard basis itself maximizes the incoherence parameter, i.e., $\max_{i,j} |\langle \mathbf{e}_i, \mathbf{e}_j \rangle|^2 = 1$.

Theorem 2 continues to hold if the rows of $\mathbf{A}$ are chosen uniformly at random from an orthonormal matrix $\mathbf{U}$ that is ν-incoherent to the standard basis. In that case, the incoherence parameter ν determines the number of measurements that are sufficient for recovering $\mathbf{x}$. Specifically, as shown by Candès and Plan (2011), ℓ_1-minimization succeeds with high probability [CP11], provided that

$$m \geq c\nu s \log n.$$

We finally remark that the aforementioned result continues to hold when the signal is sparse in an arbitrary orthonormal basis $\{\mathbf{b}_1, \ldots, \mathbf{b}_n\}$. In that case, the incoherence condition becomes $\max_{i,j} |\langle \mathbf{b}_i, \mathbf{u}_j \rangle|^2 \leq \nu/n$.

Next, suppose that $\mathbf{A}$ is a Gaussian random matrix, i.e., its entries are i.i.d. $\mathcal{N}(0, 1/m)$ distributed. Note that the rows of a Gaussian random matrix are, with high probability, incoherent with respect to any fixed orthonormal basis. We next consider the Gaussian case in more detail as it simplifies the analysis, and also because Gaussian random matrices are essentially optimal compressive sensing matrices. We hasten to add that Gaussian random matrices do typically not arise in practice, and even if we can choose the matrix $\mathbf{A}$ in an application, then there are often computationally more attractive structured random matrices with very similar properties.

We next state and prove a non-uniform recovery result for Gaussian random matrices that parallels Theorem 2. Later, we will see a strictly stronger uniform recovery result for Gaussian random matrices.

Theorem 3 *Let $\mathbf{x} \in \mathbb{R}^n$ be s-sparse, and let $\mathbf{y} = \mathbf{A}\mathbf{x}$, where $\mathbf{A}$ is a Gaussian random matrix with i.i.d. $\mathcal{N}(0, 1/m)$ entries. Fix $\delta > 0$. Then, as long as the number of measurements satisfies*

$$m \geq cs \log(n/\delta), \tag{4.10}$$

with probability at least $1 - \delta$, $\mathbf{x}$ is the unique minimizer of BP.

4.5.1 Proof of Theorem 3

The proof consists of two parts. First, we give a sufficient condition for ℓ_1-minimization to succeed, and second we show that the random matrix $\mathbf{A}$ satisfies this condition with high probability. While we assume $\mathbf{A}$ to be Gaussian, variants of the argument presented in the following also allow to treat the more structured random matrices $\mathbf{A}$ discussed at the beginning of this section. We start with the sufficient condition.

Lemma 1 *Let $\mathbf{x}$ be supported on $\mathcal{S}$. Suppose that the sub-matrix of $\mathbf{A}$ corresponding to the columns in the set $\mathcal{S}$, $\mathbf{A}_\mathcal{S}$, has full rank. If there exists a vector $\mathbf{g} = \mathbf{A}^T \boldsymbol{\nu}$ such that*

$$\mathbf{g}_\mathcal{S} = \operatorname{sign}(\mathbf{x}_\mathcal{S}) \quad \text{and} \quad \|\mathbf{g}_{\mathcal{S}^c}\|_\infty < 1, \tag{4.11}$$

then $\mathbf{x}$ is the unique minimizer of ℓ_1-minimization applied to $\mathbf{y} = \mathbf{A}\mathbf{x}$.

The vector $\mathbf{g}$ is referred to as a *dual certificate*; its existence certifies optimality; it is determined by $\boldsymbol{\nu}$ which is an optimal (not necessarily unique) solution to the dual of the ℓ_1-minimization problem. The condition in the lemma is also necessary for $\mathbf{x}$ to be a unique minimizer. Lemma 1 can be proven using duality theory, however it also follows directly from basic linear algebra, as we demonstrate next.

Proof of Lemma 1 First, suppose there exists another solution $\mathbf{x}'$ obeying $\mathbf{y} = \mathbf{A}\mathbf{x}'$ that is supported on $\mathcal{S}' \subset \mathcal{S}$. This contradicts linear independence of the columns of $\mathbf{A}_{\mathcal{S}}$. Next suppose the alternative solution is supported on $\mathcal{S}' \not\subseteq \mathcal{S}$. We show that the existence of a dual certificate implies $\|\mathbf{x}'\|_1 > \|\mathbf{x}\|_1$. We have that

$$\begin{aligned}
\|\mathbf{x}'\|_1 &> \langle \mathbf{x}', \mathbf{g} \rangle \\
&= \langle \mathbf{x}', \mathbf{g} \rangle + \|\mathbf{x}\|_1 - \langle \mathbf{x}, \mathbf{g} \rangle \\
&= \|\mathbf{x}\|_1 + \langle \mathbf{x}' - \mathbf{x}, \mathbf{g} \rangle \\
&= \|\mathbf{x}\|_1 + \boldsymbol{\nu}^T \mathbf{A}(\mathbf{x}' - \mathbf{x}) \\
&= \|\mathbf{x}\|_1.
\end{aligned}$$

Strict inequality above holds since $\mathcal{S}' \not\subseteq \mathcal{S}$ implies that there is at least one non-zero entry of $\mathbf{x}'$ with index $i \in \mathcal{S}^c$, and this non-zero multiplies with an entry of $\mathbf{g}$ that is strictly smaller than one.

Also observe that $\|\mathbf{x}'\|_1 \geq \|\mathbf{x}\|_1 + \langle \mathbf{x}' - \mathbf{x}, \mathbf{g} \rangle$ for all $\mathbf{x}'$, simply says that $\mathbf{g}$ lies in the sub-differential of $\|\cdot\|_1$ at $\mathbf{x}$. □

The difficult part of the proof is to find a certificate, and to show that it is valid. Hoping that $\|\mathbf{g}_{\mathcal{S}^c}\|_\infty$ is small, we take the certificate as the solution to the following optimization problem:

$$\underset{\boldsymbol{\nu}}{\text{minimize}}\ \|\boldsymbol{\nu}\|_2 \quad \text{subject to} \quad \mathbf{A}_{\mathcal{S}}^T \boldsymbol{\nu} = \text{sign}(\mathbf{x}_{\mathcal{S}}).$$

Provided that $\mathbf{A}_{\mathcal{S}}$ has full rank, which guarantees that this optimization problem has a solution, the solution is given in closed form:

$$\boldsymbol{\nu} = \mathbf{A}_{\mathcal{S}} (\mathbf{A}_{\mathcal{S}}^T \mathbf{A}_{\mathcal{S}})^{-1} \text{sign}(\mathbf{x}_{\mathcal{S}}).$$

We note that a certificate used to prove Theorem 2 can be constructed in the same manner. We next verify that $\boldsymbol{\nu}$ defined above obeys the conditions of Lemma 1, in particular $\|\mathbf{g}_{\mathcal{S}^c}\|_\infty < 1$.

The proof relies on the following proposition that guarantees that the sub-matrix $\mathbf{A}_{\mathcal{S}}$ is well-conditioned.

Proposition 7 *Let* $\mathbf{A} \in \mathbb{R}^{m \times n}$ *be a Gaussian random matrix. Then, for any* $\epsilon \in (0,1)$, *the event*

$$\mathcal{E}_\epsilon = \left\{ \left| \|\mathbf{A}_\mathcal{S}\mathbf{z}\|_2^2 - \|\mathbf{z}\|_2^2 \right| \le \epsilon \|\mathbf{z}\|_2^2, \text{ for all } \mathbf{z} \in \mathbb{R}^s \right\}$$

occurs with probability at least $1 - 2e^{-m\frac{\epsilon^2}{15}+4s}$.

Proposition 7 is non-trivial and follows from concentration of measure inequalities and an epsilon-net argument. We provide a proof of Proposition 7 later in this chapter.

Suppose that the event $\mathcal{E}_\epsilon$ with $\epsilon = 1/2$ occurs (the particular choice of 1/2 is arbitrary). The proposition guarantees that

$$\mathrm{P}\left[\mathcal{E}_{1/2}^c\right] \le 2e^{-m\frac{(1/2)^2}{15}+4s} \le \frac{\delta}{2},$$

where the last inequality holds by the assumption (4.10) by choosing the constant c sufficiently large.

On the event $\mathcal{E}_\epsilon$, the matrix $\mathbf{A}_\mathcal{S}$ has full rank, therefore the first condition in (4.11), i.e., $\mathbf{g}_\mathcal{S} = \mathrm{sign}(\mathbf{x}_\mathcal{S})$ holds true.

We next verify condition $\|\mathbf{g}_{\mathcal{S}^c}\|_\infty < 1$. Note that $\boldsymbol{\nu}$ only depends on $\mathbf{A}_\mathcal{S}$, therefore $\boldsymbol{\nu}$ and $\mathbf{a}_i$, $i \in \mathcal{S}^c$ are independent. It follows that, conditioned on $\boldsymbol{\nu}$,

$$\mathbf{a}_i^T\boldsymbol{\nu} \sim \mathcal{N}(0, \|\boldsymbol{\nu}\|_2^2/m), \text{ for all } i \in \mathcal{S}^c.$$

Thus, the union bound and a tail bound on a Gaussian random variable yields

$$\mathrm{P}\left[\max_{i\in\mathcal{S}^c} |\mathbf{a}_i^T\boldsymbol{\nu}| \ge t\frac{\|\boldsymbol{\nu}\|_2}{\sqrt{m}}\right] \le \sum_{i\in\mathcal{S}^c} \mathrm{P}\left[|\mathbf{a}_i^T\boldsymbol{\nu}| \ge t\frac{\|\boldsymbol{\nu}\|_2}{\sqrt{m}}\right] \le \sum_{i\in\mathcal{S}^c} 2e^{-\frac{t^2}{2}} \le 2ne^{-\frac{t^2}{2}} = \frac{\delta}{2},$$

where the last inequality follows by choosing $t = \sqrt{2\log(4n/\delta)}$.

The proof is concluded by showing that, on the event $\mathcal{E}_\epsilon$,

$$t\frac{\|\boldsymbol{\nu}\|_2}{\sqrt{m}} = 2\log(2n/\delta)\frac{\|\boldsymbol{\nu}\|_2}{\sqrt{m}} < 1. \tag{4.12}$$

To this end, note that, on the event $\mathcal{E}_{1/2}$,

$$\begin{aligned}
\|\boldsymbol{\nu}\|_2 &= \|\mathbf{A}_\mathcal{S}(\mathbf{A}_\mathcal{S}^T\mathbf{A}_\mathcal{S})^{-1}\mathrm{sign}(\mathbf{x}_\mathcal{S})\|_2 \\
&\le \|\mathbf{A}_\mathcal{S}\| \|(\mathbf{A}_\mathcal{S}^T\mathbf{A}_\mathcal{S})^{-1}\| \|\mathrm{sign}(\mathbf{x}_\mathcal{S})\|_2 \\
&\le \sqrt{1+1/2}\frac{1}{1-1/2}\sqrt{s} \\
&\le 3\sqrt{s}.
\end{aligned}$$

Thus, by choosing the constant in the Condition (4.10) sufficiently large, inequality (4.12) holds true, which shows that $\|\mathbf{g}_{\mathcal{S}^c}\|_\infty < 1$ holds with probability at least $1 - \delta/2$. By the union bound, both conditions in (4.11) hold simultaneously with probability at least $1 - \delta$, which concludes the proof.

4.6 Robustness, uniform recovery, and the restricted isometry property

Recall from Section 4.4 that ℓ_1-minimization succeeds if the columns of a matrix are incoherent. If the columns of the matrix are incoherent, sub-matrixes with few columns are close to being orthogonal. Here, we introduce the restricted isometry property (RIP) [CT06] which ensures that every s-sparse sub-matrix of $\mathbf{A}$ is near-orthogonal. This notion will turn out to be very useful to ensure that ℓ_1-minimization succeeds.

Definition 5 $\mathbf{A}$ *is said to satisfy the RIP of order s with constant* ϵ_s, *if for all s-sparse vectors* $\mathbf{x}$,

$$\left|\|\mathbf{A}\mathbf{x}\|_2^2 - \|\mathbf{x}\|_2^2\right| \leq \|\mathbf{x}\|_2^2 \epsilon_s.$$

The RIP ensures that every sub-matrix of $\mathbf{A}$ consisting of s columns is near-orthogonal, and therefore the matrix $\mathbf{A}$ preserves the length of all s-sparse vectors.

To see the connection to sparse reconstruction, note that if the RIP or order $2s$ holds, then every set of $2s$ columns is linearly independent and therefore ℓ_0-minimization provably succeeds, as we saw earlier.

More importantly, if the RIP with a sufficiently small constant holds, ℓ_1-minimization provably succeeds, as established by the following proposition, proven at the end of this section.

Theorem 4 *Suppose* $\mathbf{A}$ *satisfies the RIP of order* $2s$ *with constant* $\epsilon_{2s} < 1/3$. *Then* ℓ_1*-minimization recovers all s-sparse vectors from* $\mathbf{y} = \mathbf{A}\mathbf{x}$.

More is true. The RIP ensures that approximating a sparse vector under noise is possible, and that a variant of ℓ_1-minimization approximates a sparse vector well under noise.

Specifically, consider a noisy measurement

$$\mathbf{y} = \mathbf{A}\mathbf{x} + \mathbf{e},$$

where $\mathbf{e} \in \mathbb{R}^m$ is an additive noise term. The RIP guarantees that small amounts of noise cannot have a large effect, as there cannot exist two s-sparse vectors $\mathbf{x}, \mathbf{x}'$ that are very different but yet produce very similar measurements:

$$\|\mathbf{A}\mathbf{x} - \mathbf{A}\mathbf{x}'\|_2^2 \geq (1 - \epsilon_{2s})\|\mathbf{x} - \mathbf{x}'\|_2^2.$$

We next state (without proof) a robustness guarantee based on the RIP, which is originally due to Candès et al. (2006). [CRT06b].

Theorem 5 ([FR13, Thm. 6.11]). *Suppose* $\mathbf{A}$ *satisfies the RIP of order* $2s$ *with constant* $\epsilon_{2s} < 0.62$. *Then the following two statements hold:*

(a) For every $\mathbf{x}$, *the solution* $\hat{\mathbf{X}}$ *of* ℓ_1*-minimization applied to* $\mathbf{y} = \mathbf{A}\mathbf{x}$ *approximate the vector* $\mathbf{x}$ *in the sense that*

$$\|\hat{x} - \mathbf{x}\|_1 \leq c\sigma_s(\mathbf{x}),$$

where $\sigma_s(\mathbf{x})$ *is the best* s*-sparse approximation in* ℓ_1*-norm, i.e.,* $\sigma_s(\mathbf{x}) = \inf\{\|\mathbf{x} - \mathbf{x}_s\|_1 : \mathbf{x}_s \text{ is } s\text{–sparse}\}$.

(b) For any s*-sparse* $\mathbf{x}$ *and* $\mathbf{y} = \mathbf{A}\mathbf{x} + \mathbf{e}$ *with noise vector* $\mathbf{e}$ *obeying* $\|\mathbf{e}\|_2 \leq \eta$, *a solution to*

$$\underset{\mathbf{x}}{\textit{minimize}}\ \|\mathbf{x}\|_1 \textit{ subject to } \|\mathbf{y} - \mathbf{A}\mathbf{x}\|_2 \leq \eta \tag{4.13}$$

obeys

$$\|\mathbf{x}\|_2 - \hat{\mathbf{x}} \leq c\eta.$$

We are interested in matrices that satisfy the RIP for as large as possible values of s. A variety of random matrices satisfy the RIP. For example, the random matrix obtained by sub-sampling m columns of the Fourier matrix and scaling appropriately, and Gaussian random matrices obey the RIP with high probability provided that

$$m \geq cs\log(n/s). \tag{4.14}$$

This is formally stated in the following theorem, which is proven in the remainder of this section.

Theorem 6 *Let* $\mathbf{A} \in \mathbb{R}^{m\times n}$ *be a Gaussian random matrix with i.i.d.* $\mathcal{N}(0, 1/m)$ *entries. With probability* $1 - \delta$, $\mathbf{A}$ *obeys the RIP of order* s *with constant* ϵ_s *provided that*

$$m \geq \frac{c}{\epsilon_s^2}[s\log(n/s) + \log(1/\delta)]. \tag{4.15}$$

A consequence of this result is that provided that the number of measurements obeys (4.14), ℓ_1-norm minimization provably reconstructs an s-sparse vector (by Theorem 4), and provably approximates an s-sparse vector from a noisy observation (by Theorem 5).

This is remarkable since it establishes that with a near-linear number of measurements, the combination of random sampling and reconstruction of ℓ_1-minimization provably and computationally efficiently enables reconstruction of s-sparse vectors.

4.6.1 Proof of Theorem 4

This proof is from Foucart and Rauhut (2013) [FR13, Chapter 6.2]. We start from the recovery Condition (4.5). This recovery condition ensures recovery for a given vector $\mathbf{x}$. A popular recovery condition that guarantees that ℓ_1-minimization succeeds uniformly, that is for all s-sparse $\mathbf{x}$ is the *null-space-property*. The null-space-property characterizes when ℓ_1-minimization succeeds in recovering an s-sparse vector.

Lemma 2 *Every s-sparse vector $\mathbf{x}$ is the unique solution to ℓ_1-minimization applied to $\mathbf{y} = \mathbf{A}\mathbf{x}$ if and only if $\mathbf{A}$ satisfies the null-space-property of order s, i.e., if for all subsets $\mathcal{S} \subset [n]$ of cardinality at most s,*

$$\|\mathbf{h}_{\mathcal{S}}\|_1 < \|\mathbf{h}_{\mathcal{S}^c}\|_1 \text{ for all } \mathbf{h} \in \mathrm{Null}(\mathbf{A}).$$

Proof of Lemma 2 We only prove the "if" direction. We show that for any s-sparse vector the null-space-property implies that Condition (4.4) holds. Let $\mathcal{S}$ be the support of $\mathbf{x}$ and note that

$$\|\mathbf{x} + \mathbf{h}\|_1 = \|\mathbf{x}_{\mathcal{S}} + \mathbf{h}_{\mathcal{S}}\|_1 + \|\mathbf{h}_{\mathcal{S}^c}\|_1 \geq \|\mathbf{x}_{\mathcal{S}}\|_1 - \|\mathbf{h}_{\mathcal{S}}\|_1 + \|\mathbf{h}_{\mathcal{S}^c}\|_1 > \|\mathbf{x}\|_1,$$

by the null-space-property. This concludes the proof. □

We now show that RIP implies the null-space-property. Let $\mathcal{S}_0$ be the set of s-largest entries of $\mathbf{h}$, and $\mathcal{S}_1$ be the set of next s-largest entries of $\mathbf{h}$ and so on. Since $\mathbf{h}$ is in the kernel of $\mathbf{A}$, we have $\mathbf{A}\mathbf{h} = 0$ which can be written as $\mathbf{A}\mathbf{h}_{\mathcal{S}_0} = -\mathbf{A}(\mathbf{h}_{\mathcal{S}_0} + \mathbf{h}_{\mathcal{S}_2} + \ldots)$, where $\mathbf{h}_{\mathcal{S}}$ is the vector obtained from $\mathbf{h}$ with all entries not in $\mathcal{S}$ set to zero.

Using the RIP assumption, we have that

$$\begin{aligned}
\|\mathbf{h}_{\mathcal{S}_0}\|_2^2 &\leq \frac{1}{1-\epsilon_{2s}}\|\mathbf{A}\mathbf{h}_{\mathcal{S}_0}\|_2^2 \\
&= \frac{1}{1-\epsilon_{2s}}\langle \mathbf{A}\mathbf{h}_{\mathcal{S}_0}, -\mathbf{A}(\mathbf{h}_{\mathcal{S}_0} + \mathbf{h}_{\mathcal{S}_2} + \ldots)\rangle \\
&= \frac{1}{1-\epsilon_{2s}}\sum_{k\geq 1}\langle \mathbf{A}\mathbf{h}_{\mathcal{S}_0}, -\mathbf{A}\mathbf{h}_{\mathcal{S}_k}\rangle \\
&\overset{(i)}{\leq} \frac{1}{1-\epsilon_{2s}}\sum_{k\geq 1}\delta_{2s}\|\mathbf{h}_{\mathcal{S}_0}\|_2\|\mathbf{h}_{\mathcal{S}_k}\|_2
\end{aligned}$$

where equation (i) follows by using that $\mathbf{h}_{\mathcal{S}_0}$ and $\mathbf{h}_{\mathcal{S}_i}$ are s-sparse. Thus,

$$\begin{aligned}
\|\mathbf{h}_{\mathcal{S}_0}\|_2 &\overset{(i)}{\leq} \frac{\epsilon_{2s}}{1-\epsilon_{2s}}\sum_{k\geq 1}\|\mathbf{h}_{\mathcal{S}_k}\|_2 \\
&\leq \frac{\epsilon_{2s}}{1-\epsilon_{2s}}\sum_{k\geq 0}\frac{1}{\sqrt{s}}\|\mathbf{h}_{\mathcal{S}_k}\|_1 \\
&= \frac{\epsilon_{2s}}{1-\epsilon_{2s}}\frac{1}{\sqrt{s}}\|\mathbf{h}\|_1 \\
&\overset{(ii)}{\leq} \frac{1}{2\sqrt{s}}\|\mathbf{h}\|_1,
\end{aligned}$$

where for equation (i) we used $\|\mathbf{h}_{\mathcal{S}_k}\|_2\sqrt{s} \leq \|\mathbf{h}_{\mathcal{S}_{k-1}}\|_1$, by definition of the sets, and equation (ii) follows from the assumption $\epsilon_{2s} < 1/3$. This implies that $\|\mathbf{h}_{\mathcal{S}_0}\|_2 \leq \|\mathbf{h}\|_1$, which in turn implies the null-space-property, as desired. By Lemma 2, the null-space-property guarantees that ℓ_1-minimization recovers all s-sparse vectors from $\mathbf{y} = \mathbf{A}\mathbf{x}$.

4.6.2 Towards the proof of Theorem 6: Random projection and the RIP

Random projection is an important tool in high-dimensional data analysis. Our interest in studying random projection is the connection of random projections to the RIP.

Random projection is a technique to reduce the dimensionality of a set of points in a Euclidean space. Specifically, consider a set $\mathcal{U}$ of N points in $\mathbb{R}^n$. Suppose we want to embed these points into a lower dimensional Euclidean space (i.e., in $\mathbb{R}^m$ with $m < n$), while approximately preserving the distances between the points in $\mathcal{U}$. The Johnson-Lindenstrauss Lemma, stated below, guarantees that any set of N

points can be embedded in $m = O(\log N/\epsilon^2)$ dimensions without distorting any of the pairwise distances by more than a factor of $1 \pm \epsilon$.

Lemma 3 *(Johnson-Lindenstrauss Lemma). Choose ϵ with $0 < \epsilon < 1$ and suppose m satisfies*

$$m \geq \frac{24}{\epsilon^2} \log(2N). \tag{4.16}$$

Then, for every set $\mathcal{U}$ of N points, there is a map $f\colon \mathbb{R}^n \to \mathbb{R}^m$ such that for all $\mathbf{x}, \mathbf{x}' \in \mathcal{U}$,

$$(1 - \epsilon)\|\mathbf{x} - \mathbf{x}'\|_2^2 \leq \|f(\mathbf{x}) - f(\mathbf{x}')\|_2^2 \leq (1 + \epsilon)\|\mathbf{x} - \mathbf{x}'\|_2^2. \tag{4.17}$$

The *Johnson-Lindenstrauss* Lemma is essentially tight according to Alon (2003) [Alo03, Thm. 9.3].

The map in the *Johnson-Lindenstrauss* can be taken as a Gaussian random matrix, specifically the *Johnson-Lindenstrauss* follows directly by the following concentration inequality which guarantees that a Gaussian random matrix preserves the norm of any vector with high probability.

Lemma 4 *Let $\mathbf{A} \in \mathbb{R}^{m\times n}$ be a random matrix with i.i.d. $\mathcal{N}(0, 1/m)$ entries. Then, for any ϵ with $0 < \epsilon < 1$ and any fixed $\mathbf{x} \in \mathbb{R}^n$,*

$$\mathrm{P}\left[\left|\|\mathbf{A}\mathbf{x}\|_2^2 - \|\mathbf{x}\|_2^2\right| \geq \epsilon\|\mathbf{x}\|_2^2\right] \leq 2e^{-m\frac{\epsilon^2}{12}}. \tag{4.18}$$

Lemma 4 states that the random variable $\|\mathbf{A}\mathbf{x}\|_2^2$ is concentrated around its expectation $\mathbb{E}\left[\|\mathbf{A}\mathbf{x}\|_2^2\right] = \|\mathbf{x}\|^2$. An inequality of the form (4.18) is called "concentration of measure inequality" or simply "concentration inequality" in the literature. Remarkably, Lemma 4 continues to hold for other random matrices, such as matrices with i.i.d. sub-Gaussian entries[1], but also for more structured random matrices, albeit with a change of the constant 1/12 in (4.18).

We next briefly show that the *Johnson-Lindenstrauss* lemma is an immediate consequence of Lemma 4. A reference for this proof is Vempala (2005) [Vem05] and Dasgupta and Gupta (2003) [DG03].

Proof of the Johnson-Lindenstrauss Lemma We show that the (linear) map $f(\mathbf{x}) = \mathbf{A}\mathbf{x}$, where $\mathbf{A} \in \mathbb{R}^{m\times n}$ a Gaussian random matrix, satisfies (4.17) for all

[1] A random variable X is called sub-Gaussian if its tail probability obeys $\mathrm{P}\left[|X| > t\right] \leq c_1 e^{-c_2 t^2}$ for constants c_1, c_2.

$\mathbf{x}, \mathbf{x}' \in \mathcal{U}$ with non-zero probability. By the union bound over all $N(N-1)/2 < N^2$ pairs of points in $\mathcal{U}$, it follows from Lemma 4 that (4.17) is violated for any pair of points $(\mathbf{x}, \mathbf{x}')$ with $\mathbf{x}, \mathbf{x}' \in \mathcal{U}$ with probability less than $N^2 2e^{-m\frac{\epsilon^2}{12}}$. The proof is concluded by noting that, by Condition 4.16, $N^2 2e^{-m\frac{\epsilon^2}{12}} \leq 1/4$. □

Our interest in random projections is the following relation of random projection matrices and the RIP.

Theorem 7 *(a) Suppose that* $\mathbf{A}$ *obeys the concentration inequality (4.18). Then, with probability at least* $1 - \delta$, $\mathbf{A}$ *obeys the RIP of order s provided that*

$$m \geq \frac{c}{\epsilon^2}[s \log(n/s) + \log(1/\delta)].$$

(b) Conversely, suppose that $\mathbf{A}$ *obeys the RIP with constant* $\epsilon_{2s} \leq \eta/4$ *for* $s \geq 16 \log(4/\delta)$. *Then, for any* $\mathbf{x}$,

$$\mathrm{P}\left[\left|\|\mathbf{ADx}\|_2^2 - \|\mathbf{x}\|_2^2\right| \geq \eta \|\mathbf{x}\|_2^2\right] \leq \delta,$$

where $\mathbf{D}$ *is a matrix with random sign on its diagonal.*

Part (a) is attributed to Mendelson et al. [MPT08] and Baraniuk et al. [Bar+08], and part (b) is by Krahmer and Ward [KW11]. We prove part (a) in Section 4.6.3 below.

Together with Lemma 4, part (a) of Theorem 7 implies that Gaussian matrices obey the RIP of order $2s$ with $\epsilon < 1/3$ with high probability provided that

$$m \geq cs \log(n/s). \tag{4.19}$$

Thus, by the discussion in Section 4.6, Condition (4.19) guarantees that any s-sparse vector can be recovered perfectly from Gaussian measurements.

Proof of Lemma 4 First observe that □

$$\mathbb{E}\left[\|\mathbf{Ax}\|_2^2\right] = \mathbb{E}\left[\mathbf{x}^T\mathbf{A}^T\mathbf{Ax}\right] = \mathbf{x}^T\mathbb{E}\left[\mathbf{A}^T\mathbf{A}\right]\mathbf{x} = \mathbf{x}^T\mathbf{Ix} = \|\mathbf{x}\|_2^2.$$

Next, let $\mathbf{a}^T{}_j$ be the j-th row of $\mathbf{A}$, and set $X_j := \frac{\sqrt{k}}{\|\mathbf{x}\|_2}\mathbf{a}^T{}_j\mathbf{x}$. The random variable X_j is $\mathcal{N}(0,1)$ distributed—this follows by observing that $\mathbf{a}_j^T\mathbf{x}$ is the sum of independent Gaussians and is therefore $\mathcal{N}(0, \|\mathbf{x}\|_2^2/m)$ distributed. Next set $X = \sum_{j=1}^m X_j^2$. With this notation, we have

$$X = \sum_{j=1}^{m} X_j^2 = \frac{m}{\|\mathbf{x}\|_2^2} \sum_{j=1}^{m} |\mathbf{a}^T{}_j \mathbf{x}|^2 = \frac{m}{\|\mathbf{x}\|_2^2} \|\mathbf{Ax}\|_2^2.$$

Thus, for $\lambda \geq 0$,

$$\begin{aligned}
\mathrm{P}\left[\|\mathbf{Ax}\|_2^2 \geq (1+\epsilon)\|\mathbf{x}\|_2^2\right] &= \mathrm{P}\left[X \geq (1+\epsilon)m\right] \\
&= \mathrm{P}\left[e^{\lambda X} \geq e^{\lambda(1+\epsilon)m}\right] \\
&\leq \frac{1}{e^{(1+\epsilon)m\lambda}} \mathbb{E}\left[e^{\lambda X}\right] && (4.20) \\
&= \frac{1}{e^{(1+\epsilon)m\lambda}} \prod_{j=1}^{m} \mathbb{E}\left[e^{\lambda X_j^2}\right] && (4.21) \\
&= \frac{1}{e^{(1+\epsilon)m\lambda}} \left(\mathbb{E}\left[e^{\lambda X_1^2}\right]\right)^m, && (4.22)
\end{aligned}$$

where we used Markov's inequality for a non-negative random variable in (4.20), independence of X_j for (4.21), and that all X_j have the same distribution for (4.22).

It remains to evaluate the moment generating function $\mathbb{E}\left[e^{\lambda X_1^2}\right]$. Since X_1 is $\mathcal{N}(0,1)$ distributed,

$$\begin{aligned}
\mathbb{E}\left[e^{\lambda X_1^2}\right] &= \int_{-\infty}^{\infty} e^{\lambda x^2} \frac{1}{\sqrt{2\pi}} e^{-\frac{x^2}{2}} \, dx \\
&= \frac{1}{\sqrt{1-2\lambda}} \int_{-\infty}^{\infty} \frac{\sqrt{1-2\lambda}}{\sqrt{2\pi}} e^{-\frac{x^2}{2}(1-2\lambda)} \, dx \\
&= \frac{1}{\sqrt{1-2\lambda}}, && (4.23)
\end{aligned}$$

where we used that the integrand is the normal density with standard deviation $\frac{1}{\sqrt{1-2\lambda}}$. The conclusion above holds for any $\lambda < 1/2$. Using (4.23) in (4.22) yields

$$\mathrm{P}\left[\|\mathbf{Ax}\|_2^2 \geq (1+\epsilon)\|\mathbf{x}\|_2^2\right] \leq \left(\frac{e^{-2(1+\epsilon)\lambda}}{1-2\lambda}\right)^{\frac{k}{2}}. \tag{4.24}$$

We next minimize the RHS of (4.24). To this end, we choose λ such that the term $\frac{e^{-2(1+\epsilon)\lambda}}{1-2\lambda}$ is minimal. It is easily verified (by setting the derivative to zero) that the optimal choice is $\lambda = \frac{\epsilon}{2(1+\epsilon)}$. With this choice,

$$\mathrm{P}\left[\|\mathbf{Ax}\|_2^2 \geq (1+\epsilon)\|\mathbf{x}\|_2^2\right] \leq ((1+\epsilon)e^{-\epsilon})^{\frac{k}{2}} = \left(e^{\log(1+\epsilon)-\epsilon}\right)^{\frac{k}{2}} < e^{-\epsilon^2 \frac{k}{12}}, \tag{4.25}$$

where for the last inequality we used

$$\log(1+\epsilon) \leq \epsilon - \frac{\epsilon^2}{2} + \frac{\epsilon^3}{3} = \epsilon - \frac{\epsilon^2}{2}\left(1 - \frac{2}{3}\epsilon\right) \leq \epsilon - \frac{\epsilon^2}{6}.$$

Here, the first inequality is a consequence of the Taylor expansion of $\log(1+\epsilon)$, and the second holds for $\epsilon \leq 1$.

Similarly, we obtain

$$\mathrm{P}\left[\|\mathbf{Ax}\|_2^2 \leq (1-\epsilon)\|\mathbf{x}\|_2^2\right] \leq e^{-\epsilon^2 \frac{k}{12}}. \tag{4.26}$$

Combining (4.25) and (4.26) via the union bound concludes the proof.

4.6.3 Proof of Theorem 6: RIP for Gaussian random matrices

Theorem 6 follows from Proposition 7, restated here for convenience.

Proposition 8. *Let $\mathbf{A} \in \mathbb{R}^{m \times n}$ be a Gaussian random matrix with i.i.d. $\mathcal{N}(0, 1/m)$ entries, and let $\mathcal{S} \subset [n]$ be a fixed set with cardinality s. Then*

$$\left|\|\mathbf{Ax}\|_2^2 - \|\mathbf{x}\|_2^2\right| \leq \epsilon\|\mathbf{x}\|_2^2, \textit{for all } \mathbf{x} \in \mathbb{R}^n \textit{ supported on } \mathcal{S}, \tag{4.27}$$

holds with probability at least $1 - 2e^{-m\frac{\epsilon^2}{15}+4s}$.

Note that the concentration inequality (4.18) guarantees that the inequality (4.27) holds with high probability for any *fixed* s-sparse vector $\mathbf{x}$. Since the set of all s-sparse vector is an infinite set, we cannot simply use the union bound to conclude that (4.27) holds true. Instead, we use an ϵ-net argument: We select a subset of the unit ball of vectors supported on $\mathcal{S}$ and use the concentration inequality (4.18) and the union bound to show that, with high probability, the inequality in (4.27) holds for each point in this subset. We chose this set sufficiently dense, so we can conclude that the inequality in (4.27) continues to hold for all other vectors supported on $\mathcal{S}$. See the proof at the end of this section for details.

Theorem 6 is an immediate consequence of Proposition 7. To see this, note that, by a union bound,

$$\begin{aligned}
&\mathrm{P}\left[\left|\|\mathbf{Ax}\|_2^2 - \|\mathbf{x}\|_2^2\right| \geq \epsilon\|\mathbf{x}\|_2^2, \text{for some } s\text{–sparse } \mathbf{x}\right] \\
&\leq \sum_{\mathcal{S}\subset[n]:|\mathcal{S}|=s} \mathrm{P}\left[\left|\|\mathbf{Ax}\|_2^2 - \|\mathbf{x}\|_2^2\right| \geq \epsilon\|\mathbf{x}\|_2^2, \text{for some } \mathbf{x} \text{ supported on } \mathcal{S}\right] \\
&\leq \binom{n}{s} e^{-m\frac{\epsilon^2}{15}+4s} \\
&\leq \left(\frac{en}{s}\right)^s e^{-m\frac{\epsilon^2}{15}+4s} \\
&\leq \delta,
\end{aligned}$$

where the last inequality holds by assumption (4.15).

Proof of Proposition 8: Define the unit ball as

$$\mathcal{B} := \{\mathbf{u} \in \mathbb{R}^s : \|\mathbf{u}\|_2 \leq 1\}.$$

For $\rho \in (0, 1/2)$, there exists a subset $\mathcal{N}_\rho$ of $\mathcal{B}$ satisfying [FR13, C3]

$$\min_{\mathbf{u}\in\mathcal{N}_\rho} \|\mathbf{u} - \mathbf{z}\|_2 \leq \rho, \quad \text{for all } \mathbf{z} \in \mathcal{B},$$

of cardinality at most

$$|\mathcal{N}_\rho| \leq \left(1 + \frac{2}{\rho}\right)^s.$$

The concentration inequality (4.18) and a union bound imply that

$$\left|\|\mathbf{Ax}\|_2^2 - \|\mathbf{x}\|_2^2\right| \leq \|\mathbf{x}\|_2^2 t, \text{ for all } \mathbf{x} \in \{\mathbf{I}_{\mathcal{S}}\mathbf{u} : \mathbf{u} \in \mathcal{N}_\rho\}, \tag{4.28}$$

with probability at least

$$1 - 2\left(1 + \frac{2}{\rho}\right)^s e^{-m\frac{t^2}{12}}.$$

Here, $\mathbf{I}_{\mathcal{S}}$ is the sub-matrix of the identity consisting of rows indexed by $\mathcal{S}$, and t is chosen later. We next show that this bound on the elements of the net can be used to bound any other element of the set.

Note that (4.28) implies that

$$|\mathbf{u}^T \underbrace{(\mathbf{A}^T{}_{\mathcal{S}}\mathbf{A}_{\mathcal{S}} - \mathbf{I})}_{\mathbf{B}:=} \mathbf{u}| \leq t, \quad \text{for all } \mathbf{u} \in \mathcal{B}.$$

Now consider any vector $\mathbf{z} \in \mathcal{B}$ and choose a vector $\mathbf{u} \in \mathcal{N}_\rho$ obeying $\|\mathbf{u} - \mathbf{z}\|_2 \leq \rho$. We obtain

$$\begin{aligned} |\langle \mathbf{Bz}, \mathbf{z}\rangle| &= |\langle \mathbf{Bu}, \mathbf{u}\rangle - \langle \mathbf{B}(\mathbf{z} + \mathbf{u}), \mathbf{z} - \mathbf{u}\rangle| \\ &\leq |\langle \mathbf{Bu}, \mathbf{u}\rangle| + \|\mathbf{B}\| \|\mathbf{z} + \mathbf{u}\|_2 \|\mathbf{z} - \mathbf{u}\|_2 \\ &\leq t + \|\mathbf{B}\| 2\rho. \end{aligned}$$

Taking the maximum over all $\mathbf{z} \in \mathcal{B}$ yields $\|\mathbf{B}\| \leq t + \|\mathbf{B}\|2\rho$, which in turn is

$$\|\mathbf{B}\| \leq \frac{t}{1 - 2\rho}.$$

Setting $t = (1 - 2\rho)\epsilon$, we can conclude that

$$\mathrm{P}\left[\|\mathbf{B}\| \geq \epsilon\right] \leq 2\left(1 + \frac{2}{\rho}\right)^s e^{-m\frac{(1-2\rho)^2\epsilon^2}{12}} \leq 2e^{-m\frac{\epsilon^2}{15} + 4s},$$

where the last inequality follows by setting $\rho = 0.05$.

4.7 Chapter notes

Sparse reconstruction techniques remain important for imaging today. For example, sparse reconstruction is still widely used in medical imaging, including for accelerated magnetic resonance imaging, computed tomography, and positron emission tomography.

Sparse reconstruction techniques were also amongst the first techniques to embrace learning in computational imaging: Early works, such as the papers of Olshausen and Field [OF97], Aharon et al. [AEB06], and Elad and Aharon [EA06], proposed to learn a dictionary that can be used for image reconstruction.

As we will see later in this book, sparsity-based reconstruction methods also provide a foundation of several deep learning-based reconstruction approaches, such as unrolled neural networks.

There are several resources that provide a more detailed introduction to sparse reconstruction than this chapter, Mallat's [Mal08] book is an important reference for sparse representations, and Foucart and Rauhut [FR13] provide a comprehensive summary of the theory of sparse reconstruction.

5

Plug-and-play methods

For the reconstruction quality of an optimization-based estimator of the form as in Equation (2.1), i.e.,

$$\hat{\mathbf{x}} = \arg\min_{\mathbf{x}} \mathrm{DC}(\mathbf{x}, \mathbf{y}) + \lambda \mathrm{R}(\mathbf{x}), \tag{5.1}$$

the choice of regularizer R or signal prior is critical. Many advances in image processing methods in the past years stem from designing better image priors. However, it is difficult to come up with good signal priors, and often prior assumptions about signals cannot be easily formulated as an explicit regularizer.

Significant advances have been made over the years for denoising, where the goal is to estimate an unknown signal $\mathbf{x}$ from a noisy observation $\mathbf{y} = \mathbf{x} + \text{noise}$. Some of the best-performing denoising methods, such as BM3D [Dab+07], do not have an explicit form for the regularizer R that can be plugged into the optimization-based reconstruction approach (5.1), and thus cannot immediately be applied to solving more general inverse problems than denoising.

Plug-and-play methods enable leveraging the advances made in denoising for solving a general inverse problem, such as to recover an unknown signal from noisy linear measurements $\mathbf{y} = \mathbf{A}\mathbf{x} + \text{noise}$, by using a denoiser as a building block in an algorithm.

Plug-and-play methods iterate between a denoising step that applies a denoiser such as BM3D or a neural network and a step that improves data consistency. Only the data consistency step takes the measurement process into account, and only the denoising step takes prior assumptions about the signal into account.

The premise of plug-and-play methods is that they enable translating the performance of a denoiser to a broader class of image reconstruction methods. They perform very well for a variety of inverse problems, and are very flexible to changes in the forward model [Kam+23]. In the following, we briefly discuss plug-and-play methods.

5.1 Plug-and-play methods based on incorporating denoisers in proximal methods

Venkatakrishnan et al. [VBW13] proposed a framework for using a denoiser as a building block to solve a more general inverse problem than denoising

Deep Learning for Computational Imaging. Reinhard Heckel, Oxford University Press. © Reinhard Heckel (2025).
DOI: 10.1093/oso/9780198947172.003.0005

and coined this framework plug-and-play. This approach is based on incorporating a denoiser into an iterative optimization algorithm applied to the optimization problem (5.1), where the data consistency term is the least-squared loss, i.e.,

$$\hat{\mathbf{x}} = \arg\min_{\mathbf{x}} \underbrace{\frac{1}{2}\|\mathbf{A}\mathbf{x} - \mathbf{y}\|_2}_{\mathrm{DC}(\mathbf{x},\mathbf{y})} + \lambda \mathrm{R}(\mathbf{x}). \tag{5.2}$$

We consider two examples of this approach, one based on incorporating a denoiser into the proximal gradient descent steps for solving the optimization problem (5.2), and the original approach by [VBW13] based on incorporating a denoiser into another iterative optimization algorithm called alternating direction method of multipliers (ADMM).

5.1.1 Plug-and-play based on proximal gradient descent

Recall that proximal gradient descent is an effective algorithm for solving a convex optimization problem with a loss function consisting of a smooth and a non-smooth term. The proximal gradient descent iterations for minimizing the loss (5.2) consist of the updates

$$\mathbf{z}_{t+1} = \mathbf{x}_t - \eta \nabla \mathrm{DC}(\mathbf{x}_t, \mathbf{y}) \tag{5.3}$$

$$\mathbf{x}_{t+1} = \mathrm{prox}_{\eta\lambda\mathrm{R}}(\mathbf{z}_{t+1}), \tag{5.4}$$

where η is a step size parameter, DC is the data consistency term, and the proximal operator is defined as

$$\mathrm{prox}_{\lambda\mathrm{R}}(\mathbf{z}) = \arg\min_{\mathbf{x}} \frac{1}{2}\|\mathbf{x} - \mathbf{z}\|_2^2 + \lambda \mathrm{R}(\mathbf{x}). \tag{5.5}$$

The first update (5.3) of the variable $\mathbf{z}$ only involves the forward model and improves data consistency of the iterate. This step does not involve the regularizer.

The second update (5.4) on the variable $\mathbf{x}$ can be viewed as a denoising step that denoises the estimate $\mathbf{z}_{t+1}$. The first term in the proximal step can be interpreted as a log-likelihood term corresponding to a Gaussian density, and the second term is the regularizer; thus the proximal step can be viewed as a Gaussian denoising step which denoises the signal $\mathbf{z}$ using the regularizer R.

Often we do not have an explicit form of the regularizer, and thus we cannot solve the optimization problem (5.5) directly. Instead, we simply apply a black-box denoising algorithm D to compute the update (5.4). This denoising algorithm can be a neural network or BM3D, or another black-box denoiser. To summarize, the plug-and-play method based on proximal gradient descent computes the following updates, starting from some $\mathbf{x}^0$:

$$\mathbf{z}_{t+1} = \mathbf{x}_t - \eta \nabla \mathrm{DC}(\mathbf{x}_t, \mathbf{y}) \tag{5.6}$$

$$\mathbf{x}_{t+1} = D(\mathbf{z}_{t+1}). \tag{5.7}$$

An advantage of this formulation is that imposing data consistency via the update (5.6) does not require us to invert the forward model, we only need to compute gradients, which is advantageous when dealing with non-linear forward models.

Kamilov et al. [KMW17] introduced a version of this plug-and-play method based on accelerated proximal gradient descent. Accelerated proximal gradient descent is typically faster than proximal gradient descent (i.e., it has a faster convergence rate), and the plug-and-play variant based on accelerated proximal gradient descent also tends to converge faster than the variant based on proximal gradient descent.

5.1.2 Plug-and-play based on ADMM

Venkatakrishnan et al. [VBW13] originally designed a plug-and-play algorithm based on the iterates of ADMM algorithm. ADMM decouples the terms corresponding to the prior and the data consistency term, and similarly as in the previous paragraph, the update involving the prior is a proximal denoising step. This denoising step, performed using a black-box denoiser like BM3D or a neural network, results in the plug-and-play method. Conceptually, this idea is analogous to the plug-and-play method based on unrolling a proximal algorithm as explained above.

First note that the problem of minimizing the objective (5.2) can be written as an optimization problem minimizing two terms, each depending on a different set of variables, that are related via a linear constraint:

$$\min_{\mathbf{x},\mathbf{z}} \frac{1}{2}\|\mathbf{A}\mathbf{x} - \mathbf{y}\|_2^2 + \lambda \mathrm{R}(\mathbf{z}) \quad \text{subject to} \quad \mathbf{x} - \mathbf{z} = 0. \tag{5.8}$$

ADMM [Boy+11] applied to this objective consists of the following iterations:

$$\mathbf{x}_{t+1} = \mathrm{prox}_{\mathrm{DC}(\cdot,\mathbf{y})}(\mathbf{u}_t + \mathbf{z}_t)$$

$$\mathbf{z}_{t+1} = \mathrm{prox}_{\lambda \mathrm{R}}(\mathbf{x}_{t+1} + \mathbf{u}_t)$$

$$\mathbf{u}_{t+1} = \mathbf{u}_t + (\mathbf{x}_{t+1} - \mathbf{z}_{t+1}).$$

If we minimize jointly over the variables $(\mathbf{x}, \mathbf{z})$, this reduces to the method of multipliers. In ADMM instead, the variables $\mathbf{x}, \mathbf{z}$ are updated in an alternating fashion, which is where the name alternating direction comes from.

The update of the variable $\mathbf{x}$ involves solving a simple least-squares problem and can be solved efficiently. The second step can be interpreted as a denoising step (as discussed before) that denoises the signal $\mathbf{x}_{t+1} + \mathbf{u}_t$. Venkatakrishnan et al. [VBW13] proposed to use a black-box denoising algorithm (such as BM3D or a neural network) to perform this denoising step to update the variable $\mathbf{z}$.

ADMM and proximal gradient descent (discussed before) are guaranteed to converge to an optimal solution if the regularizer is convex. However, typically denoisers that are not convex are used for a plug-and-play method, thus plug-and-play methods are not guaranteed to converge, and do not necessarily solve an optimization problem. Current research has established several conditions under which such plug-and-play methods converge to a stationary point [KMW17]. Convergence to a stationary point does, however, not guarantee that the stationary point is close to the desired signal, and plug-and-play methods typically lack strong reconstruction guarantees that we have seen for sparse signal reconstruction in the previous chapter.

5.2 Regularization by denoising

In the previous section, we discussed a family of plug-and-play methods for solving an inverse problem by sequentially applying denoising steps, derived by unrolling proximal methods. Romano et al. [REM17] proposed an alternative framework for using a denoiser in an iterative scheme for signal reconstruction, called regularization by denoising (RED).

Romano et al. [REM17] propose a regularizer of the form

$$\mathrm{R}(\mathbf{x}) = \frac{1}{2}\mathbf{x}^T(\mathbf{x} - \mathrm{D}(\mathbf{x})),$$

where $\mathrm{D}: \mathbb{R}^n \to \mathbb{R}^n$ is a signal denoiser. The penalty induced by this regularizer is proportional to the inner product between the image itself and the residual of the denoiser applied to the image.

With this choice of regularizer, the objective (5.2) for a linear inverse problem becomes

$$\frac{1}{2}\|\mathbf{A}\mathbf{x} - \mathbf{y}\|_2^2 + \lambda\frac{1}{2}\mathbf{x}^T(\mathbf{x} - \mathrm{D}(\mathbf{x})).$$

The RED approach minimizes this objective via gradient descent, ADMM, or another iterative algorithm such as a fixed point strategy. As a concrete approach, let us use gradient descent.

Assuming that the Jacobian of the denoiser D is symmetric, the gradient of the denoiser is given by $\mathbf{x} - \mathrm{D}(\mathbf{x})$. With this assumption, RED first applies the denoiser to obtain the iterate $\mathbf{z}_t = \mathrm{D}(\mathbf{x}_t)$ and then performs a gradient descent step

$$\mathbf{x}_{t+1} = \mathbf{x}_t - \eta \left(\mathbf{A}^T(\mathbf{A}\mathbf{x}_t - \mathbf{y}) + \lambda(\mathbf{x}_t - \mathbf{z}_t) \right). \tag{5.8}$$

Similar to the plug-and-play methods based on proximal methods, this method iterates a denoising step and a step that improves data consistency.

Assuming that the Jacobian of the denoiser D is symmetric, the gradient of this denoiser is given by $\mathbf{x} - \mathrm{D}(\mathbf{x})$. However, as pointed out by Reehorst and Schniter [RS19], existing denoisers such as BM3D or a convolutional network do not have a symmetric Jacobian. Reehorst and Schniter [RS19] further show that for non-symmetric denoisers there is no regularization term $\mathrm{R}(\mathbf{x})$ that explains the RED algorithm by Romano et al. [REM17]. Nevertheless the RED algorithm performs well as a plug-and-play method.

6
Learning to solve inverse problems end-to-end

In previous chapters, we saw that algorithms for solving inverse problems can be derived by formulating an optimization problem that incorporates the measurement process and prior assumptions about the signal.

In this chapter, we discuss an entirely different approach: learning a signal reconstruction algorithm in the form of a neural network. We consider image reconstruction problems. In a nutshell, we first obtain a rough reconstruction of the image and then train a neural network to map the rough reconstruction to a clean image. The network itself is then the reconstruction algorithm, and after training it can be applied to an (unseen) measurement to reconstruct an image. This approach is called end-to-end because the entire process from the initial rough reconstruction to the final high-quality output is learned and performed by the neural network.

The motivation to learn a reconstruction method in the form of a neural network is to either improve computational speed or to improve reconstruction performance, or both, relative to the classical, optimization-based algorithms discussed earlier, such as ℓ_1-regularized least-squares.

A signal reconstructing neural network is very fast since reconstruction only requires evaluating the neural network (i.e., performing a forward pass), contrary to solving an optimization problem with an iterative optimization procedure, as required for optimization-based algorithms discussed earlier.

Recall that the key to the success of optimization-based algorithms is that the prior assumptions about the signal to be reconstructed are relatively accurate. For example, sparse signal reconstruction works well if the signal to be reconstructed is actually sparse in some basis. A learning-based approach can improve reconstruction performance over traditional hand-crafted algorithms by learning better prior assumptions about the data, and is therefore particularly interesting when assumptions about signals are difficult to formulate mathematically. A concrete example are natural images, which are only approximated by existing mathematical models.

The main subject of this chapter are neural networks trained end-to-end to perform signal reconstruction. We start with an overview of the approach, which is to learn the parameters of a neural network based on training examples. Neural networks give state-of-the-art performance for a variety of imaging problems; concrete examples are accelerated MRI and denoising. The architecture of the

Deep Learning for Computational Imaging. Reinhard Heckel, Oxford University Press. © Reinhard Heckel (2025).
DOI: 10.1093/oso/9780198947172.003.0006

network is critical for performance. Network architectures that work well take knowledge such as how the measurements are obtained and modelling assumptions on the data into account, as we will see in this and in the next chapter. In this chapter, we start with an overview of neural network architectures for imaging tasks. In particular, we discuss convolutional neural networks in length since they are the most prevalent architecture for image reconstruction tasks. Finally, we discuss the training of neural networks and how much data is required to train them.

6.1 Overview of the supervised learning approach

Consider a signal reconstruction problem, where the goal is to reconstruct a signal $\mathbf{x} \in \mathbb{R}^n$ from a measurement $\mathbf{y} \in \mathbb{R}^m$. In this chapter, we focus on images, but the general approach discussed here also works for other kinds of signals. Suppose we have access to training data consisting of N pairs of images and associated measurements:

$$\{(\mathbf{x}_1, \mathbf{y}_1), \ldots, (\mathbf{x}_N, \mathbf{y}_N)\}.$$

We assume that the measurements are generated as $\mathbf{y}_i = \mathbf{A}\mathbf{x}_i + \mathbf{e}_i$, where $\mathbf{A}$ is a measurement operator and $\mathbf{e}_i$ is noise. For example, for Gaussian denoising, the measurement matrix is the identity ($\mathbf{A} = \mathbf{I}$), and the measurements are generated as $\mathbf{y}_i = \mathbf{x}_i + \mathbf{e}_i$, where $\mathbf{e}_i$ is i.i.d. Gaussian noise. For magnetic resonance imaging, the measurement matrix $\mathbf{A}$ is a sub-sampled Fourier matrix, and $\mathbf{e}_i$ is again i.i.d. Gaussian noise.

Let $f_{\boldsymbol{\theta}}: \mathbb{R}^m \to \mathbb{R}^n$ be a neural network with parameters $\boldsymbol{\theta}$ mapping a measurement to a signal. We discuss neural network architectures below, for now it is sufficient to view the network as a function or model parameterized by the vector $\boldsymbol{\theta}$. The parameter vector $\boldsymbol{\theta}$ contains all trainable parameters of the model.

A learning-based approach to signal reconstruction is to train the neural network by minimizing the loss function

$$\mathcal{L}(\boldsymbol{\theta}) = \frac{1}{N} \sum_{i=1}^{N} \|\mathbf{x}_i - f_{\boldsymbol{\theta}}(\mathbf{y}_i)\|_2^2, \tag{6.1}$$

with an optimizer, such as gradient descent or the stochastic gradient method. The result of this training process is the parameter vector $\hat{\boldsymbol{\theta}}$. At inference, the neural network estimates a signal based on a measurement $\mathbf{y}$ as $\hat{\mathbf{x}} = f_{\hat{\boldsymbol{\theta}}}(\mathbf{y})$. This process learns a signal reconstruction algorithm in the form of a neural network.

Here, we chose the loss function between the reconstructed image, $f_{\boldsymbol{\theta}}(\mathbf{y}_i)$, and the clean image, $\mathbf{x}_i$, as the mean-squared error. The mean-squared error is often used for quantitatively evaluating the image reconstruction performance of an algorithm, and is therefore a popular choice. Moreover, the loss can be interpreted probabilistically by assuming that the residuals $\mathbf{x}_i - f_\theta(\mathbf{y}_i)$ are normally distributed with zero mean and constant variance. Then the sum (6.1) is the negative log-likelihood of the data under this Gaussian error model, up to a constant. Minimizing this sum corresponds to maximum likelihood estimation.

However, a reconstruction with a comparatively small mean-squared error is not always perceived as a visually better reconstruction, as discussed in more detail in Chapter 14. Therefore, often other loss functions are used, such as the ℓ_1-norm, or the structural similarity metric [WSB03], as discussed in Chapter 14. As long as the loss function is differentiable, which is important for being able to minimize the loss function (as we discuss in Section 6.7), it can be used in this framework.

6.1.1 Posing reconstruction as a denoising problem

Throughout this chapter, we consider linear inverse problems, where the goal is to reconstruct an image $\mathbf{x}$ from a noisy measurement $\mathbf{y} = \mathbf{A}\mathbf{x} + \mathbf{e}$, where $\mathbf{A}$ is the measurement operator, and $\mathbf{e}$ is additive noise. We assume that we are given knowledge of the measurement operator $\mathbf{A}$.

We cast image reconstruction from the measurements $\mathbf{y}$ as a denoising problem. By taking the measurement operator $\mathbf{A}$ into account, we first compute a rough reconstruction of the signal $\mathbf{x}$, for example by computing the minimum ℓ_2-norm least-squares reconstruction which can be obtained by using the pseudo-inverse of the measurement operator $\mathbf{A}$ as $\hat{\mathbf{x}}(\mathbf{y}) = \mathbf{A}^\dagger \mathbf{y}$. The minimum-norm least-squares reconstruction is a rough estimate of the image $\mathbf{x}$ and typically has a lot of artifacts, since it takes no prior assumption on the signal $\mathbf{x}$ into account. Given pairs of measurement and corresponding clean images $(\mathbf{y}, \mathbf{x})$, we train an image-to-image neural network f_θ to map the rough reconstruction $\hat{\mathbf{x}}(\mathbf{y})$ to a clean image. See Figure 6.1 for an illustration of this approach.

This approach to image reconstruction relies on a network mapping a noisy image to a clean image. In the following, we discuss network architectures that can perform such a task well.

6.1.2 Residual learning for denoising

Before we discuss image-to-image networks, we briefly comment on training neural networks for denoising, which is an important instance of a linear inverse problem discussed in this chapter.

A neural network for denoising can be trained to simply map a noisy image to a clean image. However, training the network to predict the residual image $\mathbf{x} - \mathbf{y}$

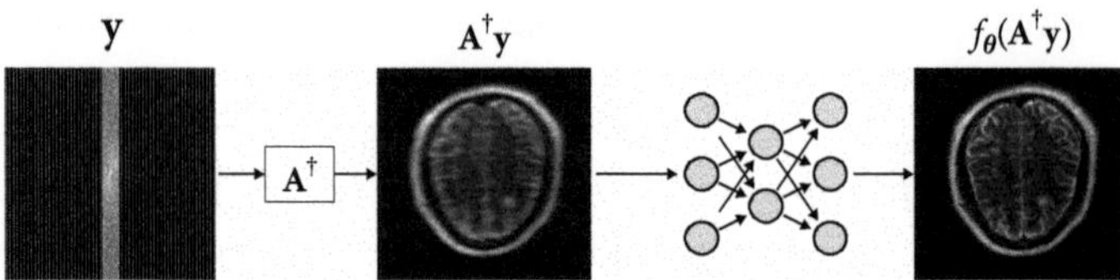

Figure 6.1 Illustration of the end-to-end reconstruction approach: First, we obtain a rough estimate of the signal by taking the measurement operator into account. An image-to-image neural network then maps the rough reconstruction to a clean image.

yields slightly better performance for Gaussian denoising, as demonstrated, for example, by Zhang et al. [Zha+17]. To perform residual learning, during training, we minimize the cost function

$$\mathcal{L}(\boldsymbol{\theta}) = \frac{1}{N} \sum_{i=1}^{N} \|\mathbf{y}_i - \mathbf{x}_i - f_{\boldsymbol{\theta}}(\mathbf{y}_i)\|_2^2.$$

At inference, we estimate the clean image by subtracting the estimated residual from the noisy observation as $\hat{\mathbf{x}} = \mathbf{y} - f_{\boldsymbol{\theta}}(\mathbf{y})$. Residual learning performs better for Gaussian denoising across architectures.

6.2 Overview of neural network architectures

A (deep) neural network uses multiple layers to map an input to an output. Neural networks perform very well for a variety of tasks including image and signal classification, prediction tasks, natural language processing, and signal and image reconstruction problems.

We view a neural network as a function parameterized by a vector of weights $\boldsymbol{\theta}$ that maps an input space $\mathbb{R}^m$ to an output space $\mathbb{R}^n$. For simplicity, we use the letter m to designate the dimension of the input space, and n to designate the dimension of the output space, but the input and output are typically images or volumes throughout. In this case, m can be interpreted as the number of measurements, and n is the number of pixels or voxels. When the output is an image, n can alternatively be a two-dimensional index representing the width and height of the image. In this chapter, the input to the network is a noisy image and the output is a clean image.

A neural network generates a prediction by transforming an input layer-by-layer into an output. The values of layer $\ell + 1$ are a function of the values in the previous layer ℓ (or of the values in the previous layers), and are obtained through linear functions and applications of non-linear functions, typically applied element-wise.

The following choices of layers, discussed in more detail later, are common in neural networks:

- **Linear layers.** The values of a linear layer are organized in a vector. The values in the $(\ell + 1)$st layer, $\mathbf{z}_{\ell+1} \in \mathbb{R}^{n_{\ell+1}}$ are given as a function of the values in the ℓ-th layer, $\mathbf{z}_\ell \in \mathbb{R}^{n_\ell}$, as

$$\mathbf{z}_{\ell+1} = \phi(\mathbf{W}_\ell \mathbf{z}_\ell + \mathbf{b}_\ell),$$

  where $\mathbf{W}_\ell \in \mathbb{R}^{n_{\ell+1} \times n_\ell}$ are the weights and $\mathbf{b}_\ell \in \mathbb{R}^{n_{\ell+1}}$ are the biases of the network. Both are trainable parameters of the layer. The function ϕ is non-linear and applied element-wise, most commonly, this is the rectangular linear unit $\mathrm{relu}(z) = \max(z, 0)$. A network consisting of only linear layers is often called a multi-layer-perceptron (MLP) or a fully-connected neural network, since each neuron in a layer depends (or is connected to) every other neuron in the previous layer.
- **Convolutional layers.** For convolutional layers, data is organized in a tensor. For images, the tensor has dimensions $w \times h \times c$, where w and h are the spacial dimensions (width and height) of an image and c is the number of channels. A convolutional layer transforms a tensor of dimensions $w_\ell \times h_\ell \times c_\ell$ to another tensor of dimensions $w_{\ell+1} \times h_{\ell+1} \times c_{\ell+1}$ by applying parameterized convolutional operations to each of the channels individually, and summing up the results. We discuss convolutional layers in detail below. Convolutional layers are one of the most important building blocks for neural networks for imaging.
- **Attention layers**. Attention layers act on a sequence of n data vectors of dimension d each, and transform those data points into another sequence of data points. For imaging applications, the data vectors are features of an image, typically they are transformed image patches. We briefly discuss attention layers below.

In the following, we discuss a few neural network architectures and how they are used for image reconstruction tasks. We will be specific as to how the networks are trained and what their parameters, such as number of layers, kernel sizes, etc. are, since those aspects are important for performance.

6.3 Denoising with a fully-connected neural network

We start with discussing a simple fully-connected neural network that works well for denoising.

Consider a denoising problem where our goal is to train a neural network f_θ to map a noisy image $\mathbf{y} = \mathbf{x} + \mathbf{e}$ to a clean image, where $\mathbf{e}$ is Gaussian noise (or noise following another distribution). We are given a training set consisting of pairs of clean and noisy images, i.e., $\{(\mathbf{x}_i, \mathbf{y}_i = \mathbf{x}_i + \mathbf{e}_i), i = 1, \ldots, N\}$. For the sake of this discussion, assume that the images have a fixed size of say 512×512 pixels, and are greyscale.

Simply training a fully-connected network f_θ consisting of a few linear layers to map a noisy image to a clean image by minimizing the loss (6.1) does not work well, for the following reasons. First, the network would be very large, since the input of the network is 512^2-dimensional, and thus the number or parameters associated only with one neuron in the second layer is 512^2. Second, the resulting network does not work well for denoising, since the network does not incorporate any prior knowledge about images in its architecture, and cannot learn sufficient prior knowledge from a reasonable number of training examples. Finally, the network would pertain to one image size, and in practice, we would like to be able to denoise images of varying sizes.

A solution to the issues described above is to split an image into overlapping image patches and denoise each patch separately with a neural network. The final image is then obtained by averaging over the overlapping patches. Burger et al. [BSH12] is an early work on image denoising with neural networks that pursued this approach. Burger et al. [BSH12] trained a four-layer fully-connected neural network to denoise image patches by minimizing the loss (6.1). The network's input is an image patch of size 17×17 reordered to a vector of size $17 * 17 = 289$, and the network's output is a vector of size 289 that is again reordered to a 17×17 image patch. The hidden layers of the network all have size 2047. The network was trained on millions of patches that were obtained from ten-thousands of images. Each training example consists of a clean image patch $\mathbf{x}_i$ and an associated noisy image $\mathbf{y}_i = \mathbf{x}_i + \mathbf{e}_i$, where $\mathbf{e}_i$ is Gaussian noise added to the image patch.

The resulting network gives similar performance to un-trained state-of-the-art denoisers at the time for Gaussian denoising (specifically to BM3D [Dab+07]), for some images yielding better, and for others worse results.

Such fully-connected networks are currently not used for image denoising applications, since architectures such as convolutional networks or attention-based networks, discussed next, perform better. However, it is interesting to see that even a simple and non-structured fully-connected network can yield good denoising performance, when applied to image patches.

6.4 Convolutional neural networks

Convolutional neural networks (CNNs for short) are a very popular architecture for vision applications in general and image reconstruction in particular. While new architectures are developed on a regular basis, at the time of writing (2023), most state-of-the-art networks for image reconstruction tasks have a convolutional architecture. This might change as new architectures, optimization algorithms, and datasets for training are developed. Networks such as the attention-based vision transformer [Dos+21], other attention-based architectures such as the Swin transformer [Lia+21], and even MLP-based architectures [MLH22] have recently emerged for image reconstruction and give sometimes slightly better results than convolutional networks; but as of now, convolutional networks continue to provide state-of-the-art or close to state-of-the-art performance, often at lower inference and training costs compared to transformer-based architectures.

In this section, we introduce convolutional neural networks, and in the next section we discuss three particular convolutional neural networks for image reconstruction tasks.

6.4.1 Convolutional operators

Convolutional neural networks apply convolutional operations, which we define first.

For simplicity, we start with the one-dimensional case, where the signal is a vector $\mathbf{x} \in \mathbb{R}^d$. A discrete convolution of $\mathbf{x}$ with the convolutional filter $\mathbf{c} = [c_1, c_2, \ldots, c_d] \in \mathbb{R}^d$ is defined as

$$[\mathbf{Cx}]_k = \sum_{i=1}^{d} x_i c_{k-i}, \tag{6.2}$$

were $[\mathbf{Cx}]_k$ is the k-th index of the vector Cx, and where the indices are taken modulo d, so $c_{-1} = c_d, c_{-2} = c_{d-1}$, etc. A convolution is a linear operation and can be written as a multiplication with a circulant matrix $\mathbf{C}$ defined as

$$\mathbf{C} = \begin{bmatrix} c_1 & c_2 & c_3 & c_4 & \cdots \\ c_{-1} & c_1 & c_2 & c_3 & \cdots \\ c_{-2} & c_{-1} & c_1 & c_2 & \cdots \\ \vdots & \vdots & \vdots & \ddots & \end{bmatrix} \in \mathbb{R}^{d \times d}. \tag{6.3}$$

In words, the rows and columns of a circular matrix are circularly shifted versions of the vector $\mathbf{c}$.

Circulant matrices are characterized by their commutativity property. Specifically, let $\mathbf{S}_i$ be the translation or shift operator that circularly shifts a vector by i entries. A matrix is circulant if and only if it commutes with shifts, i.e., if $\mathbf{S}_i\mathbf{Cx} = \mathbf{CS}_i\mathbf{x}$ for all $\mathbf{x}$ and for all shifts i. Also note that two circulant matrices $\mathbf{C}_1$ and $\mathbf{C}_2$ commute, i.e., $\mathbf{C}_1\mathbf{C}_2 = \mathbf{C}_2\mathbf{C}_1$. Another important property of circulant matrices is that they are diagonalized by the discrete Fourier transform. Specifically, let $\mathbf{F}$ be the DFT matrix, then $\mathbf{C} = \mathbf{FDF}^H$, where $\mathbf{D}$ is a diagonal matrix. This property is often used to compute a convolution efficiently, but is not used in convolutional neural networks, because convolutional neural networks have small spacial extent, so it is most efficient to compute them directly, according to the definition (6.2).

Convolutional operations for image processing are typically 2D-convolutions. The convolution of a 2D-array $\mathbf{x} \in \mathbb{R}^{d \times d}$ with a 2D-filter $\mathbf{c} \in \mathbb{R}^{d \times d}$ is defined analogously as for a one-dimensional convolution in Equation (6.2) as

$$[\mathbf{Cx}]_{k,\ell} = \sum_{i=1}^{d} \sum_{j=1}^{d} x_{i,j} c_{k-i,\ell-j}.$$

However, for convolutional neural networks, the filter $\mathbf{c}$ has typically smaller dimension than the size of the array or image $\mathbf{x}$. Common filter sizes for convolutional neural networks, discussed later, are 3×3, 5×5, and at most 11×11.

For image processing applications, it does not make sense to take the indices modulo-d, since this would presume that the array is wrapped around, like on a circle. However, we have to decide what to do around the boundary of the array. It is common to zero-pad the input array as illustrated in the 2D-convolution in Figure 6.2.

6.4.2 Convolutional layers

A convolutional neural network consists of convolutional layers. Data in a convolutional neural network for image processing is organized into a three-dimensional tensor (or array) of size $w \times d \times c$, where w and d are the spacial dimensions, and c is the number of channels. While image processing is the most common application for convolutional networks, convolutional networks are also used for one or three-dimensional data, in which case the spacial dimensions are one or three. For example, audio data has only a single spacial dimension, and a three-dimensional reconstruction of a knee in medical imaging has three spacial dimensions (width, height, and depth).

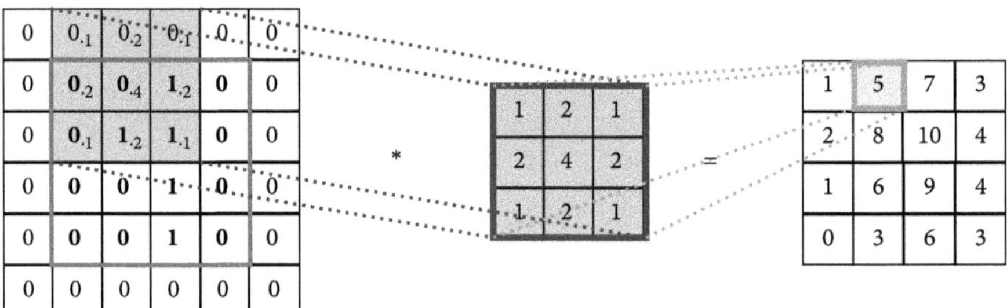

Figure 6.2 An illustration of a convolution of a 4×4 image (a course depiction of the character 1) in the orange frame with a Gaussian kernel of size 3×3. The image is zero-padded so that the convolution with a 3×3 kernel again produces an image of size 4×4. This kernel blurs the image.

A convolutional layer (with two spacial dimensions width and height) takes as input a volume $\mathbf{x}_1 \in \mathbb{R}^{w_1 \times d_1 \times c_1}$ and transforms it to an output volume $\mathbf{x}_2 \in \mathbb{R}^{w_2 \times d_2 \times c_2}$ through the transformation $\mathbf{x}_2 = \text{conv}(\mathbf{x}_1)$ defined as

$$[\text{conv}(\mathbf{x}_2)]_i = \sum_{j=1}^{c_1} \mathbf{C}_{ij}[\mathbf{x}_1]_j + b_i, \quad i = 1, ..., c_2. \tag{6.4}$$

Here, $[\text{conv}(\mathbf{x}_2)]_i \in \mathbb{R}^{w_1 \times d_1}$ is the i-th channel of the output and $[\mathbf{x}_1]_j$ is the j-th channel of the input volume. Moreover, $\mathbf{C}_{ij}$ denotes a convolutional operation with a filter $\mathbf{c}_{ij}$. For imaging applications, the convolutional filter has small spacial extent, typically the dimensions are 3×3, 5×5, or at most 11×11. Finally, b_i is a scalar bias parameter, that adds the same bias term to each element in a channel.

In words, the i-th channel of $\mathbf{x}_2$, denoted by $[\mathbf{x}_2]_i$, is computed by first convolving each of the input channels individually, and second, summing up the results. The trainable parameters of the convolutional layer are the convolutional filters and the biases. For example, a convolutional layer with 3×3 convolutions has $3^2 c_1 c_2 + c_2$ many parameters: 3^2 for each filter, and there are $c_1 c_2$ many filters, and c_2 biases, one for each output channel.

6.4.2.1 Down- and up-sampling with convolutions

The convolutional operation conv as defined in Equation (6.4) transforms a tensor into another tensor of the same dimensions. However, through so called strides the operation can also reduce or increase the dimension of the input tensor. With strides, a convolutional operation can down- or up-sample a volume. Here is how strides are defined. Consider a tensor with only one spacial dimension, $\mathbf{x} \in \mathbb{R}^d$ for simplicity. Note that a convolution of a signal $\mathbf{x}$ with a convolutional filter of spacial extent 3 can be written as a linear operator applied to the signal, $\mathbf{Cx}$, where

$$\mathbf{C} = \begin{bmatrix} c_1 & c_2 & c_3 & 0 & \dots & & \\ 0 & c_1 & c_2 & c_3 & 0\dots & & \\ 0 & 0 & c_1 & c_2 & c_3 & 0\dots \\ \vdots & \vdots & \vdots & \ddots & & \end{bmatrix} \in \mathbb{R}^{d \times d}.$$

A convolution with stride 2 and with a filter of size 3 can be written as the following linear operator:

$$\mathbf{C} = \begin{bmatrix} c_1 & c_2 & c_3 & 0 & \dots & & & & \\ 0 & 0 & c_1 & c_2 & c_3 & 0 & \dots & & \\ 0 & 0 & 0 & 0 & c_1 & c_2 & c_3 & 0 & \dots \\ \vdots & \vdots & \vdots & \ddots & & & & & \end{bmatrix} \in \mathbb{R}^{\frac{d}{2} \times d}.$$

A convolution with stride 2 reduces the spatial dimension of the output space by a factor of two relative to the input space, and thus down-samples.

A convolution with stride 1/2 and with a filter of size 3 can be written as the linear operator

$$\mathbf{C} = \begin{bmatrix} c_1 & 0 & 0 & \dots \\ c_2 & 0 & 0 & \dots \\ c_3 & c_1 & 0 & \dots \\ 0 & c_2 & 0 & \dots \\ 0 & c_3 & c_1 & \dots \\ \vdots & \vdots & \vdots & \ddots \end{bmatrix} \in \mathbb{R}^{d \times \frac{d}{2}}.$$

This operation is also called a transposed convolution. It can be viewed as a convolution with a fractional stride. The operation increases the spacial dimension of the output space by a factor of two and thus up-samples a volume. A convolution with a fractional stride is also sometimes referred to as a deconvolution, but in the context of inverse problems, deconvolution typically refers to the problem of reversing the effect of a convolution, for example to remove the distortions or blurring introduced by a convolution.

6.5 Signal reconstruction with convolutional neural networks

A convolutional neural network consists of applications of the convolutional operation conv defined in (6.4) followed by application of a non-linearity, applied element-wise. The currently most commonly used non-linearity is the rectified linear unit defined as $\text{relu}(z) = \max(z, 0)$.

We next discuss three popular example architectures for image reconstruction problems, to introduce convolutional neural networks, and to understand how convolutional networks are used for image reconstruction problems.

6.5.1 A fully-convolutional network for denoising

Convolutional neural networks were used for denoising as early as 2008, when Jain and Seung [JS08] proposed a four-layer convolutional network for Gaussian denoising and demonstrated that it could compete in image quality with the state-of-the-art approaches of the time.

A simple fully-convolutional neural network works well for denoising. Zhang et al. [Zha+17] studied a fully-convolutional network for denoising, specifically, for reconstructing a colour image $\mathbf{x} \in \mathbb{R}^{w \times h \times 3}$ from the noisy observation $\mathbf{y} = \mathbf{x} + \mathbf{e}$, where $\mathbf{e}$ is Gaussian noise. The colour image is represented as a three-dimensional tensor with three colour channels (RGB: red, green, blue). The network, illustrated in Figure 6.3, consists of 17 layers, each hidden layer has 64 channels. The first layer is transformed to a volume consisting of 64 channels by applying the conv-operation followed by a relu-non-linearity. The subsequent layers are computed by applying the conv-operation, followed by a function that normalizes channels and is discussed later (batch-normalization), followed by a relu-non-linearity. The final volume is transformed back to an image with the conv-operation. Each convolutional layer uses filters of size 3×3. The network was trained on a training set constructed based on 400 images of size 180×180. The images were split into patches of size 40×40, and noise was added to generate a pair of noisy and clean patches of the image.

The network provides a marginal improvement (about 0.5 dB PSNR) over BM3D [Dab+07], a long-standing un-trained state- of-the-art method. Today, CNN-based denoisers are considered to be state-of-the-art.

6.5.2 Super-resolution with a simple convolutional neural network

Next, we discuss a simple convolutional network that gives excellent performance at the task of super-resolution.

Recall that the super-resolution problem can be formulated as follows. We are given a low-resolution image $\mathbf{y}$ that can be viewed as obtained through blurring followed by down-sampling from a high-resolution image as $\mathbf{y} = \mathbf{DBx}$. Here, $\mathbf{B}$ is implementing blurring with a Gaussian kernel, and $\mathbf{D}$ implements a down-sampling operator. The goal is to reconstruct the high-resolution image $\mathbf{x}$ from the low-resolution observation.

Posing reconstructing as a denoising problem (discussed in Section 6.1.1), and training a simple convolutional network gives excellent super-resolution performance. Specically, Dong et al. [Don+16] proposed to first construct a rough estimate of the high-resolution image by performing bicubic interpolation on $\mathbf{y}$, and second, to train a neural network to map the rough reconstruction to a clean image.

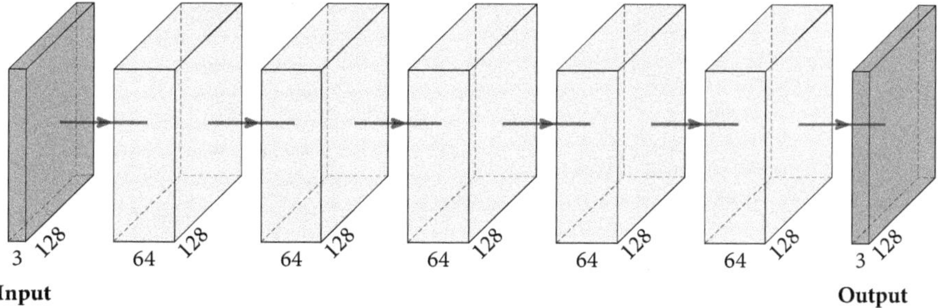

Figure 6.3 A convolutional neural network with 5 hidden layers. This network, with 17–20 hidden layers was used by Zha et al. [Zha+17] to get good denoising performance. Drawn with the PlotNeuralNet package.

The network (in its default configuration) is a three-layer convolutional network with relu-activation functions, depicted in Figure 6.4. The first layer takes the input, a colour image with three channels, one for each colour channel, and maps it to a 64-channel volume via a convolutional layer with 9×9 filters, followed by application of a relu-non-linearity. This first layer can be interpreted as generating features for each image patch. Next, the network maps those features to a volume with 32 channels, this time with 1×1 convolutional filters. Finally, the network generates an output by performing a convolution with 5×5 convolutions.

The network was trained on a training set generated from only 91 images. The images were split into non-overlapping patches or sub-images of size 33×33. Those sub-images were then used to generate pairs of low-resolution sub-image and original sub-image. The low-resolution sub-image was generated from a high-resolution image by blurring with a Gaussian kernel followed by down-sampling. The resulting network yielded state-of-the-art super-resolution performance at the time.

6.5.3 Denoising and signal reconstruction with the U-net

Next, we briefly discuss a convolutional neural network that is very popular and gives excellent performance for image reconstruction tasks. The U-net [RFB15] is an encoder-decoder architecture originally introduced for biomedical image segmentation that is a go-to for a variety of image reconstruction problems.

See Figure 6.5 for an illustration of the architecture. The encoder part of the architecture consists of the following elements: Convolutional layers with convolutional filters of size 3×3 followed by a relu-non-linearity. After two such convolutional layers, a convolutional operation with stride 2 or a pooling layer

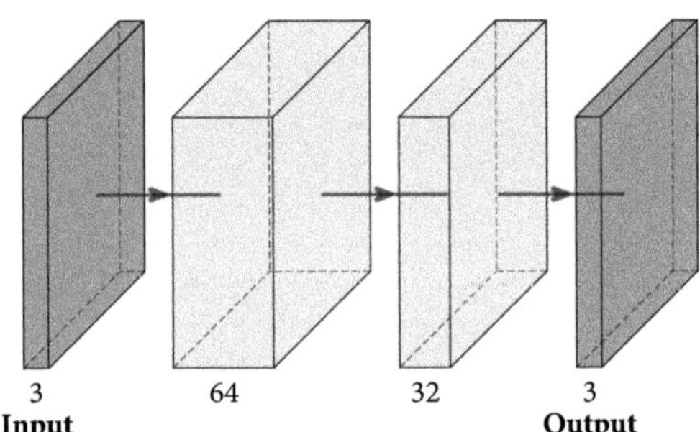

Figure 6.4 A simple three-layer convolutional network used to map a rough up-sampled image to a clean image to perform super-resolution. Drawn with the PlotNeuralNet package.

down-samples the volume by reducing the spacial dimensions by a factor of two. At the same time, the number of channels is increased by a factor of two. The decoder part of the architecture reverses those operations. It performs up-sampling with transposed convolutional layers, which up-sample the spacial dimensions by a factor of two, and decrease the number of channels by a factor of two. The network also contains skip-connections. Those skip-connections simply append the volumes of the encoder to the respective volumes of the decoder. Finally, the U-net has normalization layers that normalize each channel.

An intuition for an encoder-decoder architecture is to force the network to learn a low-dimensional representation of the image, and reconstruct the image from this low-dimensional representation. This intuition is familiar from denoising a signal $\mathbf{x}$ that lies in a low-dimensional subspace parameterized by the orthonormal basis $\mathbf{U} \in \mathbb{R}^{n \times d}$. Recall that an optimal denoiser of a signal that lies in a subspace is the projection $\mathbf{U}\mathbf{U}^T$. The operation $\mathbf{U}^T$ can be interpreted as an encoder, and the operation $\mathbf{U}$ can be interpreted as the decoder.

The encoder-decoder architecture is not critical for image-reconstruction performance; a network without up- and down-sampling layers can yield similar image quality. However, the up- and down-sampling operations make the network computationally significantly more efficient than a network such as the fully-connected network for denoising, discussed in Section 6.5.1. The reason is that the computational cost of the conv-operation scales linearly in the width, height, and number of channels of the image. Therefore, reducing the width and height by a factor of two and increasing the channels by a factor of two (as done by the U-net) reduces the computational cost of applying a conv-operation by a factor of two.

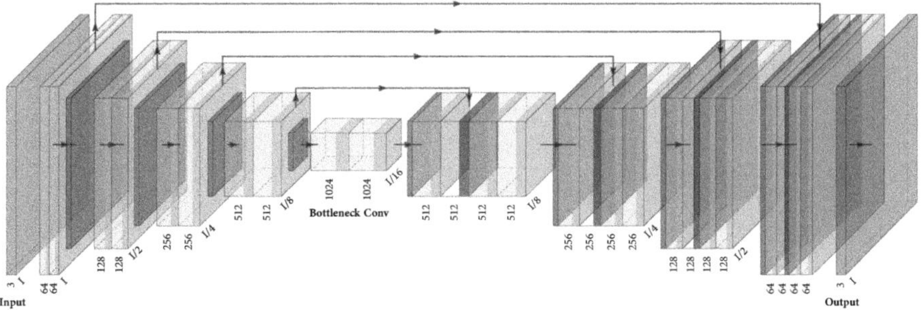

Figure 6.5 The U-net architecture particularized for mapping an image with three colour channels to another colour image with three colour channels. The network performs very well for denoising and is used as a building block in a variety of state-of-the-art signal recovery problems. Drawn with the PlotNeuralNet package.

To conclude, U-net type architectures give better denoising quality than the aforementioned fully-convolutional neural network for denoising (by about 1–2 dB in PSNR, see cf. [Bro+19, Table 1]), and are faster. Moreover, U-net architectures give excellent performance for medical image reconstruction tasks, for example for computed tomography [Jin+17] and for magnetic resonance imaging [Zbo+19].

6.6 Transformers for imaging applications

Architectures based on self-attention, in particular the transformer architecture [Vas+17] have become the standard architecture for natural language processing application. Adaptions of transformers to vision applications, in particular the vision transformer [LH21] also show excellent performance for vision applications, like image classification. Transformer architectures also work well for image reconstruction tasks, such as denoising [Che+21; LH21], in particular when pre-trained or trained on moderately large datasets. Vision transformers consist of the following attention layers.

6.6.1 Multi-head self-attention (MHSA) layers

Multi-head self-attention layers act on a sequence of n data points $\mathbf{X} = [\mathbf{x}_1, \ldots, \mathbf{x}_n] \in \mathbb{R}^{n \times d}$, each of dimension d, and typically produce a sequence of the same dimension $n \times d$. They build on self-attention layers as follows. First, for each data point, a q-dimensional query, key, and value vector is generated via a trainable linear transformation. This operation can be written in matrix form as

$$\mathbf{Q} = \mathbf{U}_Q\mathbf{X} \in \mathbb{R}^{q \times n}, \quad \mathbf{K} = \mathbf{U}_K\mathbf{X} \in \mathbb{R}^{q \times n}, \quad \mathbf{V} = \mathbf{U}_V\mathbf{X} \in \mathbb{R}^{q \times n},$$

where $\mathbf{U}_Q, \mathbf{U}_K, \mathbf{U}_V \in \mathbb{R}^{q \times d}$ are the trainable linear transformations. We then compute attention weights as

$$\mathbf{A} = \mathrm{softmax}(\mathbf{Q}^T\mathbf{K}) \in \mathbb{R}^{n \times n},$$

and finally the self-attention layers yield

$$\mathrm{SA}(\mathbf{X}) = \mathbf{A}\mathbf{V}^T \in \mathbb{R}^{n \times q}.$$

Multi-head self-attention extends self-attention by performing k independent self-attention operations (with independent trainable associated linear maps), and aggregates those maps as

$$\mathrm{MSA}(\mathbf{X}) = \mathbf{U}[\mathrm{SA}_1(\mathbf{X}), \ldots, \mathrm{SA}_k(\mathbf{X})]^T \in \mathbb{R}^{d \times n},$$

where we choose $q = d/k$ as is common, and where $\mathbf{U}$ is a linear transformation applied to the concatenated outputs of the multiple self-attention heads.

6.6.2 Vision transformer-based image-to-image architecture

A transformer architecture is a sequence-to-sequence model, i.e., it maps a sequence to a sequence. To use a transformer for an image-to-image vision application, it is critical to tokenize an image well, i.e., represent it as a sequence of tokens. A vision transformer acts on image patches. First, an image is split into a sequence of n non-overlapping image patches (for example, of size 10×10 pixels), and those image patches are mapped with a learnable linear transformation to feature vectors. To each feature vector, positional information is appended, which contain information of the absolute position of the image patches in the image. The resulting n data points of dimension d are fed through a concatenation of a few MHSA layers. A reconstruction head then maps the output of the MHSA layers to an image. The reconstruction head simply applies a linear transformation to each of the n data points, which maps them to non-overlapping image patches. Those image patches are assembled to an image.

This variant of a vision transformer performs similar to slightly better for magnetic resonance imaging than a plain U-net [LH21], and it can improve by 0.4–1.6 dB over convolutional networks for denoising and similar tasks [Che+21]. This slight improvement in performance comes at the price of a slight increase in inference and training time.

There are other effective image-to-image models that combine elements of convolutional networks and transformers that outperform the U-net slightly, such as SwinIR [Lia+21] and Restormer [Zam+22].

6.7 Optimization of neural networks

To train a neural network, we minimize a loss function of the form

$$\mathcal{L}(\boldsymbol{\theta}) = \frac{1}{N} \sum_{i=1}^{N} \mathrm{loss}(\mathbf{x}_i, f_{\boldsymbol{\theta}}(\mathbf{y}_i)), \tag{6.5}$$

where loss is for example the mean-squared error. This objective function is often called an empirical risk, since it is an empirical version of the risk

$R(f_\theta) = \mathbb{E}\left[\text{loss}(\mathbf{x}, f_\theta(\mathbf{y}))\right]$, where expectation is over a pair of jointly distributed random variables $(\mathbf{x}, \mathbf{y})$.

In principle, minimizing the empirical risk can be very difficult, since a loss involving a neural network is non-convex, and thus the loss function can have multiple local optima. However, in practice, we simply apply gradient descent (or variants thereof such as the stochastic gradient method, or adaptive gradient methods) to minimize the loss. This typically works very well, and often it even enables us to find a global optimizer, which we can identify by the empirical risk (6.1) being zero.

In order to apply gradient-based methods, we need to be able to compute gradients of the loss, which we will briefly discuss in the next subsection.

While neural networks can, in principle, be very difficult to optimize, various elements built in the architecture of the network facilitate the optimization of neural networks. We will briefly discuss those elements in the subsection about vanishing gradients below.

6.7.1 Gradient computation

In order to minimize an empirical risk, we need to compute gradients of the loss with respect to the network's parameters.

There are at least four methods for computing those derivatives. First, we could work out the gradients by hand and implement them in computer code. This is what we would do for a simple least-squares problem, for example. While working out the gradients by hand is in principle doable, it would be very tedious for practical neural networks.

Second, we could use a computer to do the tedious work for us via symbolic differentiation. Symbolic differentiation is used by software packages, such as Maple and Mathematica, and derives a symbolic expression of the gradient, similar to an expression we would derive by hand. This expression can be evaluated or implemented as a function. Symbolic differentiation results in complex expressions and is not practical for neural networks.

Third, we could approximate the derivative of a function as $f'(x) = \frac{f(x+h)-f(x-h)}{2h}$ for some small h. This is called numeric differentiation, is simple to implement, but can be very inaccurate, since we are working with floating point numbers in a computer.

Finally, there is algorithmic or automatic differentiation which is used to train neural network in practice, and is implemented by software packages, such as PyTorch, TensorFlow, and JAX. In a nutshell, automatic differentiation provides an efficient way to compute gradients of functions that can be written as a composition of differentiable functions. See Baydin et al. (2018) [Bay+18] for a survey. The software packages PyTorch, TensorFlow, and JAX provide a framework to represent neural networks as a composition of differentiable functions, and enable the efficient computation of gradients, using automatic differentiation.

We first discuss the backpropagation algorithm, and then how auto differentiation is a more general approach with a potentially lower memory footprint and higher potential for scheduling computations efficiently. The packages PyTorch, TensorFlow, and JAX implement autodifferentiation and not the backpropagation algorithm for computing derivatives.

6.7.1.1 Backpropagation algorithm

Consider a neural network $\mathbf{f}_\theta$ parameterized by the parameter vector $\boldsymbol{\theta}$. Suppose the network takes an input $\mathbf{z}_0$ and maps it to a multi-variate output $\mathbf{z}_M = \mathbf{f}_\theta(\mathbf{z}_0)$. For example, if the network outputs probability estimates corresponding to K different classes, the output is K-dimensional. We assume that the network's output is computed through M steps according to

$$\begin{aligned} \mathbf{z}_1 &= \mathbf{f}_1(\mathbf{z}_0) \\ \mathbf{z}_2 &= \mathbf{f}_2(\mathbf{z}_1) \\ &\vdots \\ \mathbf{f}_\theta(\mathbf{z}_0) = \mathbf{z}_M &= \mathbf{f}_M(\mathbf{z}_{M-1}). \end{aligned}$$

We are interested in computing the gradient $\nabla_\theta f_\theta(\mathbf{z}_0) = \nabla_\theta L(\mathbf{z}_M(\boldsymbol{\theta}))$ associated with a loss $L(\mathbf{z}_M(\boldsymbol{\theta}))$ at $\boldsymbol{\theta}$.

Define the Jacobian matrix of a function $\mathbf{f}: \mathbb{R}^n \to \mathbb{R}^m$ as the matrix containing the gradients with respect to each output coordinate as a row as

$$\mathbf{J}_\mathbf{f}(\boldsymbol{\theta}) = [\nabla f_1(\boldsymbol{\theta}), \ldots, \nabla f_m(\boldsymbol{\theta})]^T \in \mathbb{R}^{m \times n}. \tag{6.5}$$

The chain rule says

$$\mathbf{J}_{\mathbf{f}\circ\mathbf{g}}(\boldsymbol{\theta}) = \mathbf{J}_\mathbf{f}(\mathbf{g}(\boldsymbol{\theta}))\mathbf{J}_\mathbf{g}(\boldsymbol{\theta}). \tag{6.6}$$

Note that for univariate functions, this reduces to the familiar chain rule $\frac{\partial}{\partial\theta} f(g(\theta)) = f'(g(\theta))g'(\theta)$. With this, we can compute the gradient of the loss as

$$\nabla_\theta L(\mathbf{z}_M(\boldsymbol{\theta})) = \mathbf{J}_L(\mathbf{z}_M)\mathbf{J}_{\mathbf{f}_M}(\mathbf{z}_{M-1})\mathbf{J}_{\mathbf{f}_{M-1}}(\mathbf{z}_{M-2})\cdots. \tag{6.7}$$

The backpropagation algorithm computes this gradient as follows. First, we perform a forward pass to compute the quantities

$$\begin{aligned}
\text{Step 1:} \quad \mathbf{z}_1 &= \mathbf{f}_1(\mathbf{z}_0) \\
\text{Step 2:} \quad \mathbf{z}_2 &= \mathbf{f}_2(\mathbf{z}_1) \\
&\vdots \\
\text{Step M:} \quad \mathbf{z}_M &= \mathbf{f}_M(\mathbf{z}_{M-1}).
\end{aligned}$$

Second, we perform a backward pass to compute the quantities

$$\begin{aligned}
\text{Step M+1:} \quad \mathbf{J}_M &= \mathbf{J}_L(\mathbf{z}_M) \\
\text{Step M+2:} \quad \mathbf{J}_{M-1} &= \mathbf{J}_M \mathbf{J}_{\mathbf{f}_M}(\mathbf{z}_{M-1}) \\
\text{Step M+3:} \quad \mathbf{J}_{M-2} &= \mathbf{J}_{M-1} \mathbf{J}_{\mathbf{f}_{M-1}}(\mathbf{z}_{M-2}) \\
&\vdots \\
\text{Step 2M +1:} \quad \nabla_{\boldsymbol{\theta}} L(\mathbf{z}_M(\boldsymbol{\theta})) &= \mathbf{J}_1 \mathbf{J}_{\mathbf{f}_1}(\mathbf{z}_0).
\end{aligned}$$

Let us study an example for concreteness. Consider the function:

$$L(\boldsymbol{\theta}) = \frac{1}{2}\left(\sin(\mathbf{z}_0^T \boldsymbol{\theta}) - y\right)^2. \tag{6.8}$$

A forward pass consists of computing the following quantities sequentially:

$$\begin{aligned}
z_1 &= \mathbf{z}_0^T \boldsymbol{\theta} \\
z_2 &= \sin(z_1) \\
L(\boldsymbol{\theta}) &= \frac{1}{2}(z_2 - y)^2.
\end{aligned}$$

The backpropagation algorithm for this example is illustrated in Figure 6.6. A backward pass consists of performing the following computations sequentially:

$$\begin{aligned}
J_2 &= (z_2 - y) \\
J_1 &= J_2 \cos(z_1) \\
\nabla L(\boldsymbol{\theta}) &= J_1 \mathbf{z}_0.
\end{aligned}$$

6.7.1.2 Automatic differentiation

Note that to run the backpropagation algorithm, we need to keep the data produced during the forward pass (i.e., $\mathbf{z}_1, \ldots, \mathbf{z}_M$) in memory as those calculations are needed during the backward pass. In addition, it is not clear how to extend

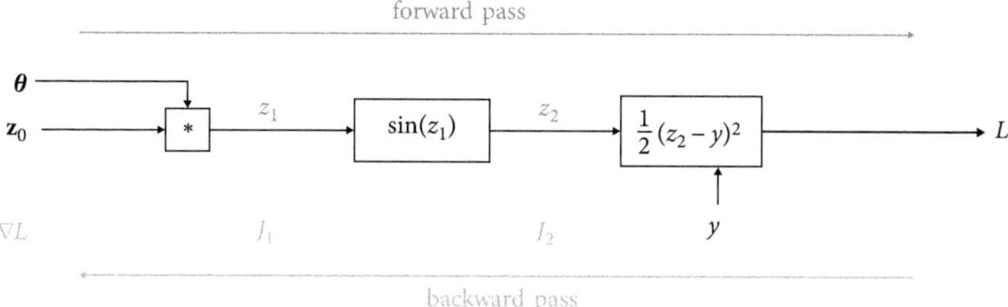

Figure 6.6 The computation graph for the function $L(\boldsymbol{\theta}) = \frac{1}{2}\left(\sin(\mathbf{z}_0^T\boldsymbol{\theta}) - y\right)^2$ and an illustration of the backpropagation algorithm for computing the gradient $\nabla L(\boldsymbol{\theta})$.

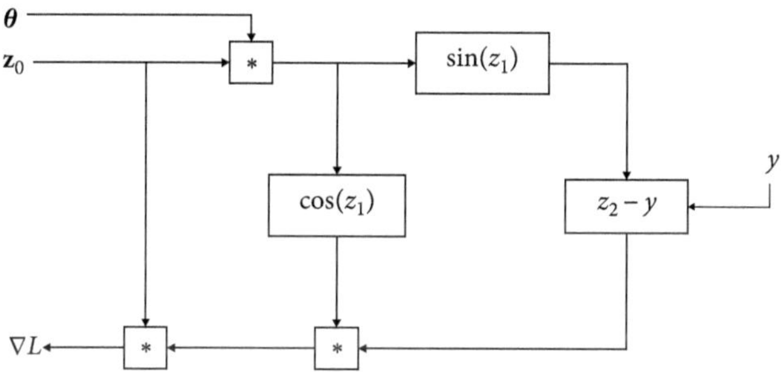

Figure 6.7 The computation graph to compute the gradient $\nabla \frac{1}{2}\left(\sin(\mathbf{z}_0^T\boldsymbol{\theta}) - y\right)^2$.

this algorithm to more complicated gradient computations, such as computing the gradient of a gradient.

To address those issues, packages like PyTorch, JAX, and TensorFlow do not run the backpropagation algorithm, instead they do something more sophisticated. They build a computation graph for gradient computation and traverse it. The computation graph expresses the entire gradient computation as a graph, and allows for better scheduling of the computation, and for a lower memory footprint. Constructing the computation graph and computing the gradient by traversing through it is called automatic differentiation.

For our example above, the computation graph is displayed in Figure 6.7.

6.7.2 Vanishing gradients

Recall that we minimize the empirical risk with gradient descent or a related first-order gradient method. The gradient-descent updates for the empirical risk (6.5) are given by

$$\boldsymbol{\theta}_{t+1} = \boldsymbol{\theta}_t - \eta \nabla \mathcal{L}(\boldsymbol{\theta}_t),$$

where η is the step size.

The neural networks we saw are a composition of linear functions and non-linearities, such as the relu-non-linearity. The gradient of such a composition of functions can vary significantly, which is problematic for the following reason. If a gradient very large, we have to choose the step size sufficiently small, so that a gradient step makes progress towards the minimum of the objective function, instead of diverging. If at the same time another gradient of the same loss function is very small, then for small gradients, a gradient update will barely move the iterate from one iteration to the other, which results in gradient descent only making very slow progress. This issue is known as the vanishing (or exploding) gradient problem. Note that the problems are the huge variations in the scale of the gradients. If the gradients were only very small or only very large, this could easily be dealt with by scaling the network's parameters or by choosing the step size accordingly.

Three popular interventions for avoiding the vanishing gradient problem are normalization layers, residual learning, and proper initialization. We discuss those approaches briefly below.

Towards this goal, let us consider a very simple version of the empirical risk minimization problem to illustrate the vanishing gradient problem, and to understand how initialization, normalization layers, and residual learning can help.

Consider the problem of minimizing the loss

$$\mathcal{L}(\theta_1, \theta_2, ..., \theta_L) = (x - y\theta_1\theta_2...\theta_L)^2,$$

where $y = 1$ with gradient descent. This is a toy example of a neural network with L layers, without non-linearities, and with a scalar input and output. We train this neural network to fit a single example. It is easy to find a solution to this optimization problem: $\theta_1 = y$ and $\theta_2 = ... = \theta_L = 1$ is a trivial solution. However, gradient descent initialized with random zero-mean Gaussian weights and with an appropriate step size takes exponentially many steps in the number of layers, L, to reach a point that is close to the optimal solution (see Shamir (2019) [Sha19] for a precise statement). This can be avoided with the following approaches.

6.7.2.1 Normalization layers

Normalization layers normalize the layers in a neural network. Popular normalization methods are batch-normalization [IS15], layer normalization [BKH16], weight normalization [SK16], and others. A simple normalization method that works well for convolutional neural network is channel normalization, which applies the following transformation to each channel $\mathbf{z}$ of a convolutional network independently

$$\mathrm{CN}(\mathbf{z}) = \frac{\mathbf{z} - \mathrm{mean}(\mathbf{z})}{\sqrt{\mathrm{var}(\mathbf{z}) + \epsilon}}\gamma + \beta, \tag{6.9}$$

where ϵ is a small scalar added for numerical stability. The parameters γ and β are channel-specific, and are either set to $\gamma = 1$ and $\beta = 0$, or are trainable. Channel normalization subtracts the mean and divides by the standard deviation, and thus standardizes the data in the channel $\mathbf{z}$.

Let us see how such a normalization helps in our toy model. Suppose we introduce the normalization operation $CN(\theta) = \theta/|\theta|$ to each layer but the last one of our toy model. This would yield the loss

$$\mathcal{L}(\theta_1, \theta_2, ..., \theta_L) = \left(x - y\theta_1 \frac{\theta_2}{|\theta_2|} \cdot ... \cdot \frac{\theta_L}{|\theta_L|}\right)^2.$$

Gradient descent applied to this loss converges linearly to an optimal solution. To see this, note that the gradients with respect to the weights $\theta_2, ..., \theta_L$ are zero, so only θ_1 is updated and, as shown previously, gradient descent applied to a quadratic converges linearly to an optimum.

In this toy model, normalization makes the deeper layers obsolete. Nevertheless this example illustrates how normalization can help the optimization. There are different suggestions for explaining the success of normalization layers, but it is generally accepted that they aid the optimization of neural network for moderately deep to deep neural networks. In shallow (two or three layer) neural networks, normalization layers are neither necessary nor used in practice.

6.7.2.2 Residual connections

Residual connections ensure that a layer is close to the identity map. In a fully-connected relu-network, the map from one layer to the other is given by $\mathbf{z}_{\ell+1} = \text{relu}(\mathbf{W}_\ell \mathbf{z}_\ell + \mathbf{b}_\ell)$. In a residual network, this map is modified by adding the input to the output:

$$\mathbf{z}_{\ell+1} = \mathbf{z}_\ell + \text{relu}(\mathbf{W}_\ell \mathbf{z}_\ell + \mathbf{b}_\ell).$$

This modification, introduced in 2015, enables to train deep neural networks (with more than 100 layers) that perform very well for image classification problems. Residual connections are often called skip-connections.

For image reconstruction problems, deep networks are not common, but skip-connections can be found in architectures like the U-net as well.

Let us consider our toy example again and see how skip-connections can help. Introducing skip-connections yield the loss

$$\mathcal{L}(\theta_1, \theta_2, ..., \theta_L) = (x - y\theta_1(1 + \theta_2) \cdot ... \cdot (1 + \theta_L))^2.$$

If we scale the parameters $\theta_2, ..., \theta_L$ to be small relative to one, then the gradients will neither be particularly large nor particularly small, thus avoiding the vanishing gradient problem.

6.7.2.3 Good initialization

The vanishing gradient descent problem can also be avoided through proper initialization, and it has been shown that very deep networks can be successfully trained without normalization layers and without residual layers, through initialization.

For example, in our toy model, if we initialize the weights $\theta_2, \ldots, \theta_L$ as 1 plus the initialization we use for the residual layers, we achieve the same effect as for residual layers for our toy example.

6.8 Scaling training set size and network size

The performance of a neural network for imaging is determined by the network architecture and optimization algorithm used, the size of the network, and the size and quality of the training set. For high-level vision tasks (such as image classification), deep networks (such as ResNets) perform well. Moreover, more data, together with scaling the network size typically improves performance, and the best networks are trained on billions of training examples. On the contrary, well-performing networks for imaging tasks can be and are often relatively shallow and are trained on relatively few training examples.

As we saw in this chapter, networks for image reconstruction tasks are typically not deep. A three-layer CNN can provide excellent super-resolution performance, and the U-net, a go-to for imaging tasks, is relatively shallow, yet yields excellent performance.

Moreover, neural networks for image reconstruction tasks require relatively few training examples compared to other machine learning tasks. Specifically, while the state-of-the-art methods for image classification and natural language processing are trained on billions of examples, the current state-of-the-art methods for image reconstruction are trained on few examples only. For example, the training set of the DnCNN model only consisted of 432 colour images, the training set of the SwinIR, a state-of-the-art transformer architecture for image denoising, only contains 10k images [Lia+21], and the most popular benchmark dataset for accelerated MRI only consists of 35k images [Zbo+19]. Often the images are split into patches and the network is trained on image patches, but still, the training set remains quite small for image reconstruction tasks.

There is evidence that scaling beyond a few thousand training images is only expected to marginally improve even the performance of large state-of-the-art models. [Don+16] trained a neural network to perform super-resolution (i.e., mapping a low-resolution image to a higher-resolution image) on only 90 example images, and observed that training on thousands of images only gives a negligible benefit (i.e., an improvement from 32.39 to 32.52 dB, which is hardly visible). In addition, only minor improvements have been observed by scaling the training set

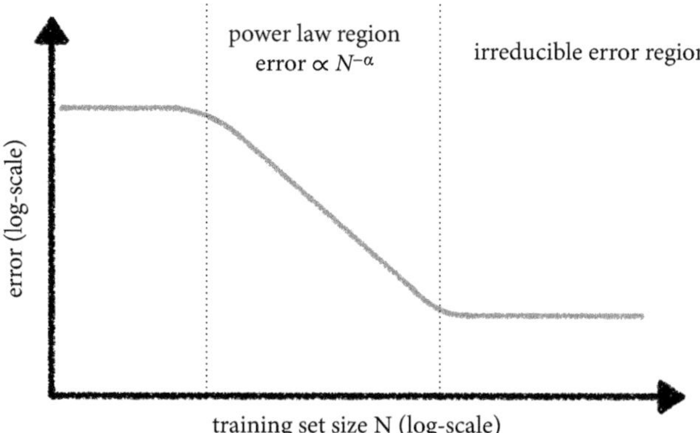

Figure 6.8 The expected qualitative behaviour of a neural network or another estimator trained on N many examples: First the error decays with the error often being well-described by a power law, and then the error reaches the irreducible error region, where increasing the training set size does not improve performance.

size for the U-net and modern transformer-based architectures, even when scaling the size of the networks along with the training set size [KH23].

6.8.1 Expected performance scaling as a function of the training examples

We next discuss the expected behaviour of the test error in more detail.

We measure performance of a neural network f_θ in terms of the risk, which is defined as the expected mean-squared error

$$R(\boldsymbol{\theta}) = \mathbb{E}_{(\mathbf{x},\mathbf{y})}\left[\|f_\theta(\mathbf{y}) - \mathbf{x}\|_2^2\right],$$

where expectation is over pairs of image and corresponding measurement $(\mathbf{x}, \mathbf{y})$ drawn from a joint distribution. We learn an estimator by minimizing the standard supervised objective function (6.1).

In general, we expect the qualitative behaviour illustrated in Figure 6.8. If the training set size is very small, then the neural network performs poorly and the risk (or the test error) is relatively large. As the training set size increases, the risk decreases as a function of the training set size, and the decrease is typically well-described by a power law. Once the training set size is sufficiently large, the risk flattens as the network reaches its irreducible error. This irreducible error region is determined by the problem, as well as the neural network used. For example, for a Gaussian denoising problem, there is a level of noise that we cannot remove as it aligns with the image manifold, this is a lower bound for the error in the irreducible error region.

6.8.2 Expected performance scaling as a function of the training examples for a simple estimator

Next we illustrate the qualitative behaviour displayed in Figure 6.8 with a concrete example. We bound the risk as a function of the number of training examples for a simple estimator and distribution that is analytically tractable.

We consider a linear denoising problem, where the joint distribution of the image and corresponding measurement is as follows. The signal $\mathbf{x}$ is drawn from a d-dimensional linear subspace according to $\mathbf{x} = \mathbf{U}\mathbf{c}$, where $\mathbf{U} \in \mathbb{R}^{n \times d}$ is an orthonormal basis for the subspace, and where $\mathbf{c} \sim \mathcal{N}(0, 1/d\mathbf{I})$. Thus, for large d, the vector is drawn approximately uniformly from the intersection of the subspace with the unit sphere. We draw a measurement as $\mathbf{y} = \mathbf{x} + \mathbf{e}$, where $\mathbf{e} \sim \mathcal{N}(0, \sigma^2/n\mathbf{I})$ is Gaussian noise. With this scaling, the expected signal-to-noise ratio (SNR) of the denoising problem is $1/\sigma^2$.

We consider a linear estimator of the form $f_{\mathbf{W}}(\mathbf{y}) = \mathbf{W}\mathbf{y}$. Note that the optimal linear estimator (i.e., the risk-minimizing estimator) is

$$\mathbf{W}_* = \frac{1}{1 + \sigma\frac{d}{n}} \mathbf{U}\mathbf{U}^T.$$

The optimal estimator projects the measurement onto the subspace and shrinks the projected measurement by a noise-dependent factor.

We provide a bound on the expected risk of the estimator that is learned by minimizing the standard supervised loss function (6.1) by running the stochastic gradient method for one epoch. Given a dataset $\mathcal{D} = \{(\mathbf{x}_1, \mathbf{y}_1), \ldots, (\mathbf{x}_N, \mathbf{y}_N)\}$ consisting of pairs of target signal and corresponding measurement drawn i.i.d. from the distribution specified above, we start the stochastic gradient method at $\mathbf{W}_0 = 0$ and update

$$\mathbf{W}_{k+1} = \mathbf{W}_k - \eta_k \nabla_{\mathbf{W}} \|\mathbf{W}\mathbf{y}_k - \mathbf{x}_k\|_2^2,$$

for $k = 1, \ldots, N$. Here, η_k is the step size.

Theorem 8 *Consider the estimate $\mathbf{W}_N$ obtained by running the stochastic gradient method for N iterations on the training set $\mathcal{D}$ with a decaying step size $\eta_k = \frac{1}{c+k}$, where c is a constant. Then the expected generalization error, where expectation is over the random training set $\mathcal{D}$ obeys*

$$\mathbb{E}\left[R(\mathbf{W}_N)\right] \leq R(\mathbf{W}_*) + 2\frac{1/d + \sigma_z^2/n}{(\sigma_z^2/n)^2}\frac{1}{N-2}. \tag{6.10}$$

The proof is based on a standard convergence analysis of the stochastic gradient method from Chapter 3 (specifically Proposition 4).

In a nutshell, the stochastic gradients we use are unbiased estimates of the gradient of the risk, $\nabla R(\boldsymbol{\theta})$. The variance of the stochastic gradient determines the rate on the right-hand side of Equation (6.10).

Theorem 8 establishes that the risk is upper bounded by a problem-dependent noise floor, which is the risk of the optimal estimator, $R(\mathbf{W}_*)$, plus a term associated with minimizing the supervised loss constructed from finitely many training examples. This term decays as c/N, which is exactly the rate we see in the simulations below for this setup.

6.9 Chapter notes

An image reconstruction algorithm in the form of a neural network trained in a supervised manner performs excellently. On popular image reconstruction benchmarks, the winning solution is typically a neural network trained in a supervised manner. For example, in the 2020 fastMRI competition [Muc+21], a competition and benchmark for 2D accelerated MRI, the best-performing methods fall into this category. As another example, for image denoising, the Smartphone Image Denoising Dataset [ALB18] and the Darmstadt Denoising Dataset [PR17] are common benchmarks, and the best-performing methods on those benchmarks are neural networks trained in a supervised manner.

Having said this, those benchmarks, and many other benchmarks, are also set up in a way that supervised methods perform well: They consist of training and test sets, and are for problems where we have pairs of measurement and target (ground-truth) images. However, in many situations of practical interest, pairs of ground-truth image and associated measurement for training are not available.

Regarding computational performance, neural networks trained in a supervised fashion often yield efficient algorithms. To reconstruct an image, we only need to perform a forward pass through a neural network which can efficiently run on a GPU. Comparatively, optimization-based methods (e.g., ℓ_1-regularized least-squares) are often much slower as they need to carry out many iterations of an iterative solver for an optimization problem.

7

Unrolled neural networks

In the previous chapter, we discussed neural networks trained in a supervised fashion to reconstruct an image. For image reconstruction, an architecture that first computes a rough estimate of the image (for example, via the pseudo-inverse of the forward map), and then maps the rough estimate to a clean image with an image-to-image network works well.

In this chapter, we discuss a technique for designing neural network architectures that typically performs even better than mapping a rough estimate to a clean image with an image-to-image network. This technique is based on modifying existing, iterative algorithms and is called unrolling or unfolding. The resulting networks iteratively refine an estimate of an image in several blocks of a neural network.

Unrolled networks are very successful architectures for image reconstruction. Currently, the best neural networks trained in a supervised fashion for accelerated magnetic resonance imaging (MRI) and computed tomography are unrolled networks. This is testified by the winning approaches of the fastMRI competition [Muc+21], a benchmark for 2D accelerated MRI reconstruction, being unrolled networks.

We again consider a linear inverse problem, where the goal is to reconstruct a signal $\mathbf{x}$ from the measurement $\mathbf{y} = \mathbf{A}\mathbf{x} + \mathbf{z}$, where $\mathbf{A} \in \mathbb{R}^{m \times n}$ is typically a wide matrix, as before, or a matrix that is poorly conditioned, and $\mathbf{z}$ is noise.

7.1 Unrolling regularized least squares

As discussed in Chapter 2, it is common to reconstruct a signal by solving a regularized least-squares problem

$$\min \mathcal{L}(\mathbf{x}), \quad \mathcal{L}(\mathbf{x}) = \frac{1}{2}\|\mathbf{A}\mathbf{x} - \mathbf{y}\|_2^2 + \mathrm{R}(\mathbf{x}), \tag{7.1}$$

where R is a regularizer.

In order to solve the regularized least-squares problem, we can use gradient descent. The gradient descent iterations starting from $\mathbf{x}_0$ with step size η_t at iteration t are

Deep Learning for Computational Imaging. Reinhard Heckel, Oxford University Press. © Reinhard Heckel (2025).
DOI: 10.1093/oso/9780198947172.003.0007

$$\begin{aligned}\mathbf{x}_{t+1} &= \mathbf{x}_t - \eta_t \nabla \mathcal{L}(\mathbf{x}_t) \\ &= \mathbf{x}_t - \eta_t \mathbf{A}^T(\mathbf{A}\mathbf{x}_t - \mathbf{y}) - \eta_t \nabla \mathrm{R}(\mathbf{x}_t).\end{aligned}$$

The first part of the gradient, $\mathbf{A}^T(\mathbf{A}\mathbf{x}_t - \mathbf{y})$, is a data-consistency term that moves the iterate towards being more consistent with the measurement.

Taking a Bayesian estimation perspective, the second part of the gradient, $\nabla \mathrm{R}(\mathbf{x}_t)$ moves the iterate towards being more likely according to the prior assumptions about the signal. Specifically, recall the Bayesian estimation perspective from Section 2.1, where the regularizer is the log-prior, i.e., $\mathrm{R}(\mathbf{x}) = \log p(\mathbf{x})$. Then the gradient of the log-prior is $\nabla \mathrm{R}(\mathbf{x}) = \nabla \log p(\mathbf{x})$, which is the score function of the distribution with density p.

The score function measures how the log-probability changes with $\mathbf{x}$; for high-probability regions (according to the signal prior) the gradient is small, and for low-probability regions the gradient is large. By subtracting the score function at point $\mathbf{x}_t$ from the image, we increase its probability under the prior distribution.

If we view the estimate $\mathbf{x}_t$ as a noisy version of the signal, then subtracting the score function $\nabla \log p(\mathbf{x})$ can be viewed as reducing noise. This makes the score function $\nabla \log p(\mathbf{x})$ a noise estimator guiding the gradient descent iterations to denoise the image.

The computation carried out by the gradient of the regularizer, $\eta_t \nabla \mathrm{R}(\mathbf{x}_t)$, can be parameterized by a neural network. For imaging tasks, we parameterize those computations with an image-to-image network, often a U-net. For T such unrolled iterations, this yields the neural network displayed in Figure 7.1. The network is initialized with a rough estimate of the image, for example $\mathbf{x}_0 = \mathbf{A}^\dagger \mathbf{y}$.

This unrolled network is then trained in a standard supervised fashion as described in the previous chapter on pairs of measurements and target images.

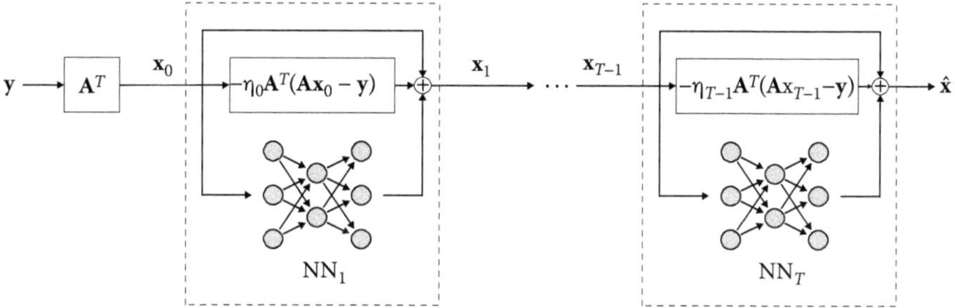

Figure 7.1 A neural network consisting of T blocks obtained by unrolling the gradient descent iterations of regularized least squares. Here, the computational step carried out by the gradients of the regularizer, $\eta_t \nabla \mathrm{R}(\mathbf{x}_t)$, is parameterized by a neural network. The network is trained in a standard supervised fashion, and the trainable parameters are the parameters of the neural networks.

Note that only the parameters of the neural networks parameterizing the gradient of the regularizer are trained, the data consistency step, $\mathbf{A}^T(\mathbf{A}\mathbf{x}_t - \mathbf{y})$, is fixed.

7.2 Overview of popular unrolled networks

In the previous section, we discussed an unrolled network obtained by unrolling the gradient iterations of regularized least squares. Other unrolled architectures can be obtained by unrolling other iterative algorithms, in fact one of the earliest unrolled networks, specifically the work of Gregor and LeCun [GL10] on learning a fast algorithm for reconstructing a sparse signal, is obtained by unrolling the iterative shrinkage thresholding algorithm (ISTA); see Section 7.3 below.

For image reconstruction, an important early work is the so-called variational network proposed by Chen et al. [CYP15] and Hammernik et al. [Ham+18] discussed in Section 7.4. The variational network is an unrolled neural network inspired by unrolling total-variation-regularized least squares. This network works well for accelerated MRI, has relatively few parameters, and is a parameter-efficient approach to signal reconstruction. The name variational network stems from the incorporation of variational principle, i.e., the formulation of the reconstruction process as an optimization problem, and unrolling this variational formulation of a signal reconstruction problem.

The network discussed in the previous section with a U-net as the neural network gives close to state-of-the-art performance for 2D accelerated MRI reconstruction (see the results for VNU in Sriram et al. (2020) [Sri+20] for performance evaluations). The corresponding network has $T = 12$ blocks and about 30M parameters. Tweaking the network further with MRI-specific steps gives another slight performance boost (see [Sri+20]) and gives state-of-the-art performance as measured in the 2020 fastMRI challenge, a benchmark for accelerated MRI reconstruction with deep neural networks [Kno+20].

In Figure 7.2 are illustrative comparison of ℓ_1-regularized least squares, U-net-based reconstruction, and Variational Network (VarNet) based reconstruction. The methods are tuned/trained on the fastMRI brain training set, and are evaluated on the fastMRI brain test-set.

Such an unrolled variational network with a U-net as the CNN also gives state-of-the-art performance for computed tomography [Gen+22].

There is a variety of other optimization algorithms and optimization problems that can be unrolled to obtain well-performing unrolled architectures, Schlemper et al. (2018) [Sch+18] is one example for MRI. Also, the image-to-image network does not have to be a U-net; other good image-to-image networks such as transformers also work well [FTS22].

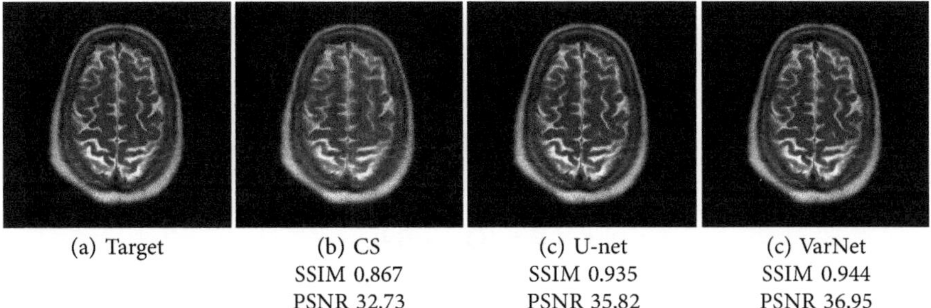

(a) Target	(b) CS SSIM 0.867 PSNR 32.73	(c) U-net SSIM 0.935 PSNR 35.82	(c) VarNet SSIM 0.944 PSNR 36.95

Figure 7.2 Reconstruction examples for reconstruction with ℓ-regularized least-squares (CS), reconstruction with the U-net as discussed in the prior chapter, and reconstruction with a network obtained by unrolling least squares and using the U-net to parameterize the gradient of the regularizer, known as a Variational Network (VarNet). It can be seen that VarNet > U-net > CS, both by visual inspection, as well as in common metrics like SSIM and PSNR. Those metrics are discussed in Section 14.3 (Measuring the similarity of two images).

7.3 Unrolling the iterative soft-thresholding algorithm

One of the earliest unrolled networks is the work of Gregor and LeCun [GL10] on learning a fast algorithm for reconstructing a sparse signal. The motivation of Gregor and LeCun [GL10] was to learn an algorithm in the form of a neural network that performs sparse reconstruction faster than a classical iterative algorithm.

Consider the ℓ_1-regularized least-squares optimization problem for reconstructing a sparse vector from a measurement $\mathbf{y}$:

$$\min_{\mathbf{x}} \frac{1}{2}\|\mathbf{A}\mathbf{x} - \mathbf{y}\|_2^2 + \lambda\|\mathbf{x}\|_1.$$

Recall that ISTA is a fast iterative algorithm for solving this optimization problem. ISTA is initialized with $\mathbf{x}_0 = 0$, and its iterations are given by

$$\mathbf{x}_{t+1} = \tau_{\lambda\eta}(\mathbf{x}_t - \eta\mathbf{A}^T(\mathbf{A}\mathbf{x}_t - \mathbf{y})),$$

where

$$\tau_\lambda(z) = \begin{cases} z + \lambda, & \text{if } z < -\lambda, \\ 0, & \text{if } z \in [-\lambda, \lambda], \\ z - \lambda, & \text{if } z > \lambda. \end{cases}$$

is the soft-thresholding operator defined earlier in Equation (3.8). The soft-thresholding operator is applied entrywise.

The motivation of the work of Gregor and LeCun [GL10] is to accelerate the existing ISTA through training. Thus, the goal is to learn a faster algorithm, but not necessarily to learn a better-performing algorithm.

We next formulate the ISTA as a neural network. The ISTA-iterates have the form

$$\mathbf{x}_{t+1} = \tau_\theta(\mathbf{Q}\mathbf{x}_t + \mathbf{B}\mathbf{y}),$$

with $\mathbf{Q} = (\mathbf{I} - \eta\mathbf{A}^T\mathbf{A})$, $\mathbf{B} = \eta\mathbf{A}^T$, and $\theta = \lambda\eta$. Suppose we carry out T many ISTA iterations. This corresponds to performing a forward-pass through a recurrent neural network, with fixed weights. To see this, we view the computations in the t-th layer of the network as applying a linear operation to the results from the previous layer, $\mathbf{x}_t$, followed by application of a non-linearity (i.e., the soft-thresholding function).

The idea of Gregor and LeCun [GL10] is to view the final result after T iterations, $f_\theta(\mathbf{y}) = \mathbf{x}_T$, as a function of the parameters $\boldsymbol{\theta} = \{\mathbf{Q}, \mathbf{B}, \theta\}$, and train those parameters on a given dataset by minimizing the standard supervised training loss (see Equation (6.1) in the previous chapter). This algorithm is called LISTA for Learned ISTA, and the network is illustrated in Figure 7.3.

In their original paper, Gregor and LeCun [GL10] find that LISTA offers a significant speedup over LISTA for few iterations. For example, a LISTA network of depth $T = 7$ reaches prediction of similar quality than ISTA run with about 10 times the number of iterations for sparse reconstruction. However, the performance tends to stagnate with the depth, and if we are willing to carry out many iterations, ISTA outperforms LISTA for sparse reconstruction.

Even in the regime where LISTA works well there is a caveat: For real-world imaging problems the size of the matrices that are learned is often prohibitively large. For example, suppose we apply this approach to magnetic resonance imaging, where the images have size say 320 × 320 pixels. This gives a parameter matrix $\mathbf{Q}$ of size $320^2 \times 320^2$, which contains about 10^{10} entries. This matrix does not fit in the memory of modern GPUs, because it would require close to 300GB in high-precision float numbers. The computational and storage burden can be decreased by parameterizing the matrix $\mathbf{Q}$ as $\mathbf{Q} = \mathbf{U}\mathbf{V}^T$, but Gregor and LeCun [GL10] found that this compromises accuracy. Even if the computational burden can be dealt

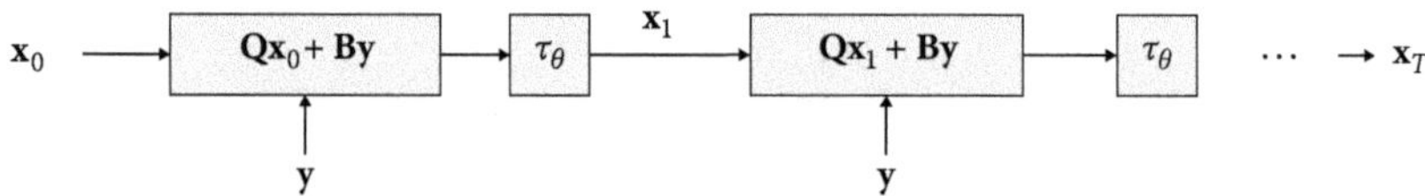

Figure 7.3 The LISTA network, the trainable parameters of the network are the matrices $\mathbf{Q}, \mathbf{B}$, and the scalar θ, and the inputs are the initial iterate $\mathbf{x}_0$ and the measurement $\mathbf{y}$.

with, note that this algorithm does not take knowledge of the forward model, **A**, into account, so the network has to learn both the signal and the forward model which is intuitively inefficient because we know the forward model.

The work of Gregor and LeCun [GL10] pertains to sparse signal recovery, but the idea of unrolling a proximal gradient descent algorithm can be applied to general iterative algorithms for solving a variety of problems, and is a useful tool for designing neural networks for signal and image recovery, as we will see in the next section.

7.4 A parameter-efficient variational network

We next discuss the so-called variational network proposed by Chen et al. [CYP15] and Hammernik et al. [Ham+18], which is an unrolled neural network inspired by unrolling total-variation-regularized least squares. This network works well for accelerated MRI, has relatively few parameters, and is therefore a parameter-efficient approach to signal reconstruction.

As discussed previously, it is common to reconstruct a signal by solving a regularized least-squares problem (7.1), i.e.,

$$\min_{\mathbf{x}} \frac{1}{2}\|\mathbf{A}\mathbf{x} - \mathbf{y}\|_2^2 + \mathrm{R}(\mathbf{x}),$$

where R is a regularizer that often induces sparsity in an appropriate domain. A popular choice for the regularizer for imaging problems is the total-variation norm, which works well if an image consists of patches that are relatively constant. This turns out to be a reasonably good model for natural images. The total-variation norm for a vector is

$$\|\mathbf{x}\|_{TV} = \sum_{i=1}^{n-1} |x_i - x_{i+1}|.$$

The total-variation norm for a vector can be written as $\|\mathbf{x}\|_{TV} = \|\mathbf{C}\mathbf{x}\|_1$, where **C** is a circular matrix with the first column equal to $\mathbf{c} = [1, -1, 0, \ldots, 0]$. In other words, the total-variation norm can be viewed as convolution with the filter $[-1, 1]$ followed by applying the non-linearity $|\cdot|$ element-wise, followed by summing up the entries:

$$\|\mathbf{x}\|_{TV} = \langle |\mathbf{C}\mathbf{x}|, \mathbf{1} \rangle,$$

where $\mathbf{1} = [1, 1, \ldots, 1]$ is the all-ones vector.

A generalization of the TV-norm regularizer is the so-called Fields of Expert model [RB09] defined as

$$\mathrm{R}(\mathbf{x}) = \sum_{i=1}^{k} \langle \psi_i(\mathbf{C}_i\mathbf{x}), \mathbf{1} \rangle,$$

where the operators $\mathbf{C}_i \colon \mathbb{R}^n \to \mathbb{R}^n$ are circulant matrices implementing convolutions with the filters $\mathbf{c}_i$ of size s. For example, the filter previously used for total-variation minimization is $[-1, 1]$, and a filter of size s is simply $[c_1, c_2, \ldots, c_s]$, where the c_j's are the filter parameters. Moreover, $\psi_i \colon \mathbb{R} \to \mathbb{R}$ is a non-linear scalar function, applied element-wise to a vector.

The original Fields of Expert model [RB09] parameterizes the non-linear activation function, similarly as the soft-thresholding function in LISTA is parameterized by the soft-thresholding parameter. Specifically, the activation function is parameterized as a linear superposition of Gaussian radial basis functions (RBFs)

$$\psi_i(z) = \sum_{j=1}^{31} w_{ij} e^{-\frac{(z-\mu_j)^2}{2\sigma^2}},$$

where μ_j are means distributed equidistant in some appropriate range, and the weights w_{ij} are the parameters of the function ψ_i.

The Fields of Experts model is parameterized by the convolutional filters $\mathbf{c}_1, \ldots, \mathbf{c}_k$ as well as the weights of the corresponding non-linearities $w_{ij}, i = 1, \ldots, k, j = 1, \ldots, 31$. Those parameters can be trained to yield an effective regularizer for image reconstruction. When originally proposed, the Fields of Experts model was used for denoising by training the parameters of the model for denoising. The Fields of Experts model can be a better regularizer for natural images than the total-variation norm, since it can express more complex regularizers than the total-variation norm.

The variational network is obtained by unrolling the gradient descent iterations of a regularized least-squares problem with a Fields of Expert regularizer. Gradient descent applied to the Fields-of-Expert regularized least-squares problem (7.1) is

$$\mathbf{x}_{t+1} = \mathbf{x}_t - \eta \left(\mathbf{A}^T(\mathbf{A}\mathbf{x}_t - \mathbf{y}) + \sum_{i=1}^{k} \mathbf{C}_i^T \psi_i'(\mathbf{C}_i\mathbf{x}_t) \right),$$

where ψ_i' is the derivative of ψ_i. Chen et al. [CYP15] proposed to unroll or unfold those iterations for a few iterates and make all parameters independently trainable, i,e, the filters in the first, second, etc. iterations are all independent of each other and are not shared.

The iterates are initialized with the rough estimate $\mathbf{x}_0 = \mathbf{A}^T\mathbf{y}$, and the remaining iterates are given by

$$\mathbf{x}_{t+1} = \mathbf{x}_t - \eta_t \mathbf{A}^T(\mathbf{A}\mathbf{x}_t - \mathbf{y}) + \mathrm{CNN}_t(\mathbf{x}_t), \tag{7.2}$$

where

$$\mathrm{CNN}_t(\mathbf{x}) = \sum_{i=1}^{k} \mathbf{C}_{t,i}^T \psi'_{t,i}(\mathbf{C}_{t,i}\mathbf{x}) \tag{7.3}$$

can be viewed as a two-layer CNN, where the trainable parameters are the convolutional filters $\mathbf{c}_{t,i}$ and the parameters of the activation function $\psi'_{t,i}$. Note that the parameters in each of the unrolled iterations are trainable, and the step size parameter η_t is trainable as well. The network is illustrated in Figure 7.1.

Chen et al. [CYP15] applied this approach successfully to denoising and JPEG deblocking. Analogously, Hammernik et al. [Ham+18] proposed such a network for magnetic resonance imaging. The network is both faster than total variation minimization, and has better imaging performance (i.e., yields better image quality).

Note that the variational network takes the forward model explicitly into account, and it also takes prior intuition about the data into account, by expressing the regularizer in a particular form inspired by the total-variation norm. The paper by Hammernik et al. [Ham+18] trained a variational network consisting of $T = 10$ steps. In each step, $k = 48$ filters of size 11×11 were learned. The activation function was parameterized by 31 RBFs. This resulted in a network with about $1.3e5$ parameters, which is a relatively small and parameter-efficient network. For example, the variational network proposed by Sriram et al. [Sri+20] using a U-net with $T = 12$ blocks and about 30M parameters, is by a factor 230 larger than the parameter-efficient variational network.

8
Self-supervised learning

Supervised training as discussed in Chapter 6 requires training data consisting of pairs of target images and associated noisy measurements. A neural network f_θ is trained on a set of such examples, $\{(\mathbf{x}_1, \mathbf{y}_1), \ldots, (\mathbf{x}_N, \mathbf{y}_N)\}$, to predict the target image based on the noisy measurement by minimizing the supervised loss

$$\mathcal{L}(\boldsymbol{\theta}) = \frac{1}{N} \sum_{i=1}^{N} \ell(f_{\boldsymbol{\theta}}(\mathbf{y}_i), \mathbf{x}_i), \tag{8.1}$$

where ℓ is a loss function such as the MSE or the ℓ_1-norm.

For some applications, it is possible to collect data consisting of target image and associated measurements. For example, for Gaussian denoising, we can generate noisy measurements from clean images by corrupting the clean images with noise. For some accelerated MRI setups, we can collect all frequencies of a given image, compute a target image from the fully-sampled measurement, and generate an undersampled measurement by retrospectively undersampling the measurement.

However, in many real-world applications, clean ground-truth images are not available, which motivates so-called self-supervised approaches that enable the training of neural networks without clean target images, for example based on noisy measurements of images only.

In this chapter, we discuss self-supervised learning approaches for denoising and for more general linear inverse problems. We focus on methods that are based on constructing a self-supervised loss for training a neural network to map a measurement to a clean image. The self-supervised loss only depends on noisy measurement data and does not require pairs of measurement and clean image.

We discuss three approaches for constructing a self-supervised loss: The first approach is based on constructing an unbiased estimate of the supervised loss (Section 8.1), which requires extra measurements but is otherwise fairly general in that it does not rely on any additional assumptions. With sufficient training data, this approach yields a network that can perform as well as the same network trained in a supervised manner.

The second approach (Section 8.3) uses Stein's unbiased estimator to construct a self-supervised loss. This enables training a network as well as training a network in a supervised manner, provided that certain assumptions are satisfied.

Deep Learning for Computational Imaging. Reinhard Heckel, Oxford University Press. © Reinhard Heckel (2025).
DOI: 10.1093/oso/9780198947172.003.0008

The third approach (Section 8.4) constructs a self-supervised loss based on predicting one part of an image from another part, and as such makes explicit or implicit assumptions about images. Even with a lot of training data, this approach typically performs worse than training a network in a supervised manner.

Self-supervised learning techniques, as discussed here, are also sometimes referred to as unsupervised training methods, since they do not require pairs of measurement and clean image. We refer to the methods considered here as self-supervised since they leverage the data or assumptions to create a loss that supervises the learning process. We reserve the term unsupervised learning for situations where an algorithm learns without explicit supervision or examples, such as when we learn a generative model from images (see Chapter 9).

8.1 Self-supervised learning based on estimates of the supervised loss

We start with a class of self-supervised methods based on constructing estimates of the supervised loss.

Consider a general inverse problem where our goal is to estimate a signal $\mathbf{x} \in \mathbb{R}^n$ from a measurement $\mathbf{y} \in \mathbb{R}^m$. Assume that a signal and an associated measurement $(\mathbf{x}, \mathbf{y})$ follow a joint distribution; for example, the image $\mathbf{x}$ could be drawn from a signal manifold and the measurement is obtained as $\mathbf{y} = \mathbf{x} + \mathbf{e}$, where $\mathbf{e}$ is Gaussian noise. This induces a joint distribution on pairs of signal and associated measurement. In this framework, our goal is to find an estimator f_{θ} that minimizes the risk

$$R(\boldsymbol{\theta}) = \mathbb{E}\left[\ell(f_{\boldsymbol{\theta}}(\mathbf{y}), \mathbf{x})\right]. \tag{8.2}$$

Here, expectation is over the signal-measurement pairs $(\mathbf{x}, \mathbf{y})$, and ℓ is a supervised loss. Note that minimizing the supervised loss requires ground-truth images $\mathbf{x}$.

Since the data distribution is unknown, we cannot compute and minimize the risk. Instead, we typically approximate the risk with the associated empirical risk (8.1), which is computed by evaluating a supervised loss on N pairs of training examples. This requires pairs of ground-truth signal and associated measurements $(\mathbf{x}_1, \mathbf{y}_1), \ldots, (\mathbf{x}_N, \mathbf{y}_N)$.

In this section, we discuss an approach to approximate the risk based on a self-supervised loss computed based on the training data. The self-supervised loss, unlike the supervised loss, does not depend on the ground-truth $\mathbf{x}$, instead it depends on another, randomized measurement of the ground-truth image, that we denote by $\mathbf{y}'$. We assume we are given N pairs of a randomized measurement and original measurement $(\mathbf{y}'_1, \mathbf{y}_1), \ldots, (\mathbf{y}'_N, \mathbf{y}_N)$ (specific choices are discussed below) and train a network to reconstruct by minimizing the self-supervised loss function

$$\mathcal{L}_{SS}(\boldsymbol{\theta}) = \frac{1}{N}\sum_{i=1}^{N} \ell_{SS}(f_{\boldsymbol{\theta}}(\mathbf{y}_i), \mathbf{y}'_i). \tag{8.3}$$

Here, ℓ_{SS} is a loss that only depends on the network's output, $f_{\boldsymbol{\theta}}(\mathbf{y}_i)$, and the randomized measurement of the ground-truth image.

We will see two examples of pairs of measurements $(\mathbf{y}, \mathbf{y}')$ and associated loss functions ℓ_{SS} that have the property that in expectation over the training data, a minimizer of the self-supervised loss is also a minimizer of the risk $R(\boldsymbol{\theta})$. The first example is for denoising and the second for compressive sensing in the context of MRI. Thus, with sufficiently many training examples we expect the same performance as with supervised training.

We will also see that the 'noise' induced by stochastically approximating the risk with the self-supervised loss (8.3) is in general larger than that induced by approximating the risk with the supervised loss (8.1), so we typically need more training examples when working with the self-supervised loss to achieve the same performance as with supervised training.

This class of methods includes the popular noise2noise methods [Leh+18]. Noise2noise methods are based on the observation that the training target of the network, which is the clean image $\mathbf{x}$ in a supervised-learning setup, can in some setups be substituted with a noisy observation of the clean image without changing the network that is learned in expectation over the training data. For denoising, the network is trained to map one noisy measurement of an image to another noisy measurement of the same image, therefore the name noise2noise.

8.1.1 Denoising

Suppose our goal is to learn a denoiser $f_{\boldsymbol{\theta}}$ that estimates an image $\mathbf{x}$ from a noisy observation $\mathbf{y} = \mathbf{x} + \mathbf{e}$, where $\mathbf{e}$ is additive noise. The additive noise can be Gaussian, but it could also be also be non-Gaussian or even structured noise. We do not have access to target images $\mathbf{x}$ for training, but we assume access to pairs of noisy observations $(\mathbf{y}_1, \mathbf{y}'_1), \ldots (\mathbf{y}_N, \mathbf{y}'_N)$, where $\mathbf{y}_i = \mathbf{x}_i + \mathbf{e}_i$ and $\mathbf{y}'_i = \mathbf{x}_i + \mathbf{e}'_i$. Here, $\mathbf{e}'_i$ is zero-mean noise, independent of the noise $\mathbf{e}_i$ and the signal $\mathbf{x}$, and not necessarily of the same distribution as the noise $\mathbf{e}_i$.

Consider the (noise2noise) self-supervised loss

$$\ell_{SS}(f_{\boldsymbol{\theta}}(\mathbf{y}), \mathbf{y}') = \|f_{\boldsymbol{\theta}}(\mathbf{y}) - \mathbf{y}'\|_2^2. \tag{8.4}$$

This loss, introduced by Lehtinen et al. [Leh+18], aims to find a network that predicts one noisy observation of an image based on another, therefore it is called noise2noise [Leh+18].

To train a network with the noise2noise loss, we need access to training data that consists of pairs of noisy measurements. An example setup, where it is possible to obtain two noisy versions of a single image is low-light photography. In low-light photography we can split an exposure into two shorter exposures, which yields a pair of noisy images of the same scene.

The self-supervised loss 8.4 is quite intuitive: The loss is small if $f_\theta(\mathbf{y})$ provides a good estimate of the measurement $\mathbf{y}'$. The noise $\mathbf{e}'$ added to the signal $\mathbf{x}$ to obtain the measurement $\mathbf{y}' = \mathbf{x} + \mathbf{e}'$ is zero-mean and independent of the signal $\mathbf{x}$, thus intuitively the best estimate of $\mathbf{y}'$ based on $\mathbf{y}$ is an estimate of the mean of $\mathbf{y}'$, which is the signal $\mathbf{x}$. If $f_\theta(\mathbf{y})$ is a good estimate of the signal $\mathbf{x}$, then the supervised loss is also small.

More formally, the rationale behind the noise2noise self-supervised loss in Equation (8.4) is that in expectation, a minimizer of the self-supervised loss is also a minimizer of the associated risk (8.2) (with ℓ equal to the mean-squared-error). This is formalized by the following statement.

Proposition 9 *Suppose that a signal* $\mathbf{x}$ *and a corresponding measurement* $\mathbf{y}$ *are drawn from a joint distribution, and let* $\mathbf{y}' = \mathbf{x} + \mathbf{e}'$ *be another randomized measurement of the signal. Assume that the noise* $\mathbf{e}'$ *is uncorrelated with the residual, i.e.,*

$$\mathbb{E}_{(\mathbf{x},\mathbf{y},\mathbf{e}')}\left[(f_\theta(\mathbf{y}) - \mathbf{x})^T \mathbf{e}'\right] = 0, \quad \textit{for all } \boldsymbol{\theta}. \tag{8.5}$$

Then the minimizer $\boldsymbol{\theta}$ *of the self-supervised risk* $\mathbb{E}_{(\mathbf{x},\mathbf{y},\mathbf{y}')}\left[\|f_\theta(\mathbf{y}) - \mathbf{y}'\|_2^2\right]$ *is also a minimizer of the supervised risk* $\mathbb{E}_{(\mathbf{x},\mathbf{y})}\left[\| f_\theta(\mathbf{y}) - \mathbf{x}\|_2^2\right]$.

To see that the proposition applies to the denoising setup above, note that any noise $\mathbf{e}'$ with zero-mean entries and independent of $\mathbf{x}$ and $\mathbf{y}$ satisfies the correlation condition (8.5). The proposition implies that for the denoising setup, training with the self-supervised loss (8.4) with infinitely many training examples gives a network that minimizes the risk. From this we expect noise2noise self-supervised training with sufficiently many training examples to yield a network that is as good as the same network trained in a supervised manner (with possibly fewer examples), and that is also what we observe in practice.

Proof of Proposition 9: We have that

$$\begin{aligned}\mathbb{E}_{\mathbf{x},\mathbf{y},\mathbf{y}'}\left[\|f_\theta(\mathbf{y})-\mathbf{y}'\|_2^2\right] &= \mathbb{E}_{\mathbf{x},\mathbf{y},\mathbf{e}'}\left[\|f_\theta(\mathbf{y})-\mathbf{x}+\mathbf{e}'\|_2^2\right]\\ &= \mathbb{E}_{\mathbf{x},\mathbf{y}}\left[\|f_\theta(\mathbf{y})-\mathbf{x}\|_2^2\right]\\ &\quad + 2\mathbb{E}_{\mathbf{x},\mathbf{y},\mathbf{e}'}\left[(f_\theta(\mathbf{y})-\mathbf{x})^T\mathbf{e}'\right] + \mathbb{E}_{\mathbf{e}'}\left[\|\mathbf{e}'\|_2^2\right]\\ &\overset{\text{(i)}}{=} \mathbb{E}_{\mathbf{x},\mathbf{y}}\left[\|f_\theta(\mathbf{y})-\mathbf{x}\|_2^2\right] + \mathbb{E}_{\mathbf{e}'}\left[\|\mathbf{e}'\|_2^2\right],\end{aligned}$$

where equation (i) follows from Condition (8.5). Noting that the expectation over the norm of the random variable $\mathbf{e}'$ above is constant as a function of $\boldsymbol{\theta}$ proves the claim. □

8.1.2 Compressive sensing

Next, we discuss a self-supervised loss for accelerated MRI. Our goal is to train a network f_θ for accelerated MRI, i.e., to reconstruct an image $\mathbf{x}$ from an undersampled measurement $\mathbf{y} = \mathbf{MFx}$, where $\mathbf{M} \in \mathbb{R}^{n\times n}$ is an undersampling mask (i.e., a diagonal matrix with zeros and ones on its diagonal), and $\mathbf{F} \in \mathbb{C}^{n\times n}$ is the Fourier transform.

Again, we do not have access to a ground-truth image $\mathbf{x}$. Instead, we are only given an undersampled measurement $\mathbf{y} = \mathbf{MFx}$ as well as a second randomized measurement of the same image, obtained as $\mathbf{y}' = \mathbf{M}'\mathbf{Fx}$, where $\mathbf{M}' \in \mathbb{R}^{n\times n}$ is a randomized mask, which is zero or one on its diagonal, and is one with non-zero probability, so that the inverse $\mathbb{E}\left[\mathbf{M}'\right]^{-1}$ exists.

From this measurement, we can evaluate a self-supervised loss defined as

$$\ell_{SS}(f_\theta(\mathbf{y}),\mathbf{y}') = \|\mathbf{W}(\mathbf{M}'\mathbf{F}f_\theta(\mathbf{y})-\mathbf{y}')\|_2^2, \tag{8.6}$$

where $\mathbf{W} = (\mathbb{E}\left[\mathbf{M}'\right])^{-1/2}$ is a diagonal weighting mask. The loss is evaluated only on the frequencies that are given through the randomized mask $\mathbf{M}'$ (note that the rows of $\mathbf{M}'\mathbf{F}f_\theta(\mathbf{y}) - \mathbf{y}'$ that are zero'd out by the mask $\mathbf{M}'$ are zero), and is weighted by how often a frequency occurs in the randomized mask $\mathbf{M}'$ (thus, the multiplication with the weighting mask $\mathbf{W}$).

Like for the noise2noise loss for denoising, the rationale behind the self-supervised loss (8.6) is that in expectation, a minimizer of the self-supervised loss is also a minimizer of the associated risk, as formalized by the following proposition.

Proposition 10 *Suppose that a signal* $\mathbf{x}$ *and a corresponding measurement* $\mathbf{y}$ *are drawn from some joint distribution (e.g., as* $\mathbf{y} = \mathbf{MFx}$*, where* $\mathbf{M}$ *is a mask), and let* $\mathbf{y}' = \mathbf{M}'\mathbf{Fx}$ *be an independent measurement taken with a randomized mask* $\mathbf{M}'$*. Then a minimizer* $\boldsymbol{\theta}$ *of the self-supervised risk* $\mathbb{E}_{\mathbf{y},\mathbf{y}'} \ell_{\mathrm{SS}}(f_\theta(\mathbf{y}), \mathbf{y}') = \mathbb{E}\left[\|\mathbf{W}(\mathbf{M}'\mathbf{F}f_\theta(\mathbf{y}) - \mathbf{y}')\|_2^2\right]$ *is also a minimizer of the supervised risk* $\mathbb{E}_{(\mathbf{x},\mathbf{y})}\left[\|f_\theta(\mathbf{y}) - \mathbf{x}\|_2^2\right]$.

Proof of Proposition 10: To see this, note that

$$
\begin{aligned}
\mathbb{E}_{\mathbf{M}'}\left[\|\mathbf{W}(\mathbf{M}'\mathbf{F}f_\theta(\mathbf{y}) - \mathbf{y}')\|_2^2\right] &= \mathbb{E}_{\mathbf{M}'}\left[\|\mathbf{W}(\mathbf{M}'\mathbf{F}f_\theta(\mathbf{y}) - \mathbf{M}'\mathbf{Fx})\|_2^2\right] \\
&= \mathbb{E}_{\mathbf{M}'}\left[\|\mathbf{WM}'\mathbf{F}(f_\theta(\mathbf{y}) - \mathbf{x})\|_2^2\right] \\
&\overset{(i)}{=} \mathbb{E}_{\mathbf{M}'}\left[\sum_j w_j^2 m_j^2 [\mathbf{F}(f_\theta(\mathbf{y}) - \mathbf{x})]_j^2\right] \\
&= \sum_j w_j^2\, \mathbb{E}_{m_j}[m_j^2][\mathbf{F}(f_\theta(\mathbf{y}) - \mathbf{x})]_j^2 \\
&\overset{(ii)}{=} \sum_j [\mathbf{F}(f_\theta(\mathbf{y}) - \mathbf{x})]_j^2 \\
&= \|f_\theta(\mathbf{y}) - \mathbf{x}\|_2^2,
\end{aligned}
$$

where equation (i) follows from $\mathbf{M}'$ and $\mathbf{W}$ being diagonal matrices, and equation (ii) follows from $\mathbf{W} = \mathbb{E}[\mathbf{M}']^{-1/2}$, which implies that $\mathbb{E}_{m_j}\left[m_j^2\right] = 1/w_j^2$. □

In practice, it is often possible to obtain such a pair $(\mathbf{y}, \mathbf{y}')$ of undersampled measurements. For example, suppose that we are given 3-times undersampled MRI measurements, obtained through undersampling with a random mask. Then we can split each measurement into a pair $(\mathbf{y}, \mathbf{y}')$, where $\mathbf{y}$ is a 4-times undersampled measurement, and $\mathbf{y}'$ is a 12-times undersampled measurement of the same image. We can now train a network for 4-times undersampled MRI on such pairs of measurements by minimizing the self-supervised loss (8.6).

Several works have demonstrated that training a neural network by minimizing a self-supervised loss of the form as in (8.6) can give excellent performance in practice [Yam+20; Zho+23]. With enough training data such self-supervised training gives essentially the same performance as supervised training. This is not surprising given that the self-supervised loss is an approximation of the risk, and the approximation error goes to zero as the number of training examples N goes to infinity.

8.1.2.1 A (bad) alternative self-supervised loss for accelerated MRI

The noise2noise self-supervised loss for denoising is based on training a network to predict a noisy image, therefore the name noise to noise. It is also possible to formulate such a loss for MRI as we discuss here. This loss, however, has a larger variance as the self-supervised loss (8.6), and is therefore suboptimal. We state it to explain that there are sometimes different ways to construct a self-supervised loss based on the same data and for a given application it is worth to identify a good one.

Given the measurement $\mathbf{y}' = \mathbf{M}'\mathbf{F}\mathbf{x}$, we can construct an estimate of the image as $\tilde{\mathbf{y}} = \mathbf{F}^{-1}\mathbf{W}\mathbf{y}'$, where $\mathbf{W} = \mathbb{E}[\mathbf{M}']^{-1}$. We can view this estimate as a noisy version of the image, with noise that is structured and not independent across pixels. Based on this, we can define a self-supervised loss analogous to the noise2noise self-supervised loss for denoising as

$$\ell_{\text{SS}}(f_\theta(\mathbf{y}), \mathbf{y}') = \|f_\theta(\mathbf{y}) - \mathbf{F}^{-1}\mathbf{W}\mathbf{y}'\|_2^2 \tag{8.7}$$

The rationale behind the self-supervised loss (8.7) is the same as for the denoising setup studied before: We train a network to predict a target that in expectation is equal to the ground-truth image.

As before, a minimizer of the self-supervised loss $\mathbb{E}_{(\mathbf{x},\mathbf{y}),\mathbf{y}'}\left[\|f_\theta(\mathbf{y}) - \mathbf{F}^{-1}\mathbf{W}\mathbf{y}'\|_2^2\right]$, is also a minimizer of the corresponding supervised risk $\mathbb{E}\left[\|f_\theta(\mathbf{y}) - \mathbf{x}\|_2^2\right]$, which follows from Proposition 9. To see this, first note that we can decompose the measurement $\tilde{\mathbf{y}}$ into the original signal plus a zero-mean noise term as

$$\tilde{\mathbf{y}} = \mathbf{F}^{-1}\mathbf{W}\mathbf{M}'\mathbf{F}\mathbf{x} = \mathbf{x} + \mathbf{e}', \quad \mathbf{e}' = \mathbf{F}^{-1}(\mathbf{W}\mathbf{M}' - \mathbf{I})\mathbf{F}\mathbf{x}.$$

The noise $\mathbf{e}'$ has zero-mean since $\mathbb{E}[\mathbf{M}'] = \mathbf{I}$. Moreover, the noise satisfies the correlation condition (8.5), thus the conditions of Proposition 9 are satisfied.

8.2 Finite-sample analysis

In the previous two sections, we saw how self-supervised losses can be constructed for denoising and compressive sensing so that in expectation, a minimizer of the self-supervised loss is also a minimizer of the associated risk. Thus, with enough training data, a network fitted with a self-supervised loss is expected to perform as well as the same network trained with a supervised loss. In practice, the amount of training data that is required increases in the variance of the self-supervised loss. Here, we formalize this intuition by analysing a simple denoising problem analytically.

We define the joint distribution on the image and corresponding measurement as follows. We assume that the signal $\mathbf{x}$ is drawn from a d-dimensional linear subspace according to $\mathbf{x} = \mathbf{U}\mathbf{c}$, where $\mathbf{U} \in \mathbb{R}^{n \times d}$ is an orthonormal basis for the subspace, and where $\mathbf{c} \sim \mathcal{N}(0, 1/d\mathbf{I})$. Thus, for larger d, the vector is drawn approximately uniformly from the intersection of the subspace with the unit sphere. We then draw a corresponding measurement as $\mathbf{y} = \mathbf{x} + \mathbf{z}$, where $\mathbf{z} \sim \mathcal{N}(0, \sigma_z^2/n\mathbf{I})$ is

Gaussian noise. With this scaling, the expected SNR of the denoising problem is $1/\sigma_z^2$.

We consider a linear estimator of the form $f_{\mathbf{W}}(\mathbf{y}) = \mathbf{W}\mathbf{y}$. Note that the optimal linear estimator (i.e., the estimator that minimizes the risk) is $\mathbf{W}^* = \frac{1}{1+\sigma_z^2\frac{d}{n}}\mathbf{U}\mathbf{U}^T$. The estimator projects the measurement onto the subspace and then shrinks the projected measurement by a noise-dependent factor.

We now provide a bound on the expected risk of the estimator that is learned by minimizing the self-supervised loss

$$\mathcal{L}_{\mathrm{SS}}(\mathbf{W}) = \frac{1}{N}\sum_{i=1}^{N} \|f_{\mathbf{W}}(\mathbf{y}_i) - \mathbf{y}_i'\|_2^2$$

by running the stochastic gradient descent method for one epoch. Specifically, we are given a dataset $\mathcal{D} = \{(\mathbf{y}_1, \mathbf{y}_1'), \ldots, (\mathbf{y}_N, \mathbf{y}_N')\}$ consisting of pairs of two noisy measurements $\mathbf{y}_i = \mathbf{x}_i + \mathbf{z}_i, \mathbf{y}_i' = \mathbf{x}_i + \mathbf{e}_i$, drawn i.i.d. from the distribution specified above, and we start the stochastic gradient method at $\mathbf{W}_0 = 0$ and update $\mathbf{W}_{k+1} = \mathbf{W}_k - \eta_k \nabla_{\mathbf{W}} \|\mathbf{W}\mathbf{y}_k - \mathbf{y}_k'\|_2^2$, for $k = 1, \ldots, N$. Here, η_k is the step size.

Theorem 9 *Consider the estimate $\mathbf{W}_N$ obtained by running the SGM for N iterations on the training set $\mathcal{D}$ with a decaying step size $\eta_k = \frac{1}{c+k}$, where c is a constant. Then the expected generalization error, where expectation is over the random training set $\mathcal{D}$ obeys*

$$\mathbb{E}\left[R(\mathbf{W}_N)\right] \leq R(\mathbf{W}^*) + \frac{1/d + \sigma_z^2/n}{(\sigma_z^2/n)^2}\frac{1}{N-2}\left(2 + \left(12\sigma_z^2\frac{d}{n} + \sigma_e^2(1+\sigma_z^2)\right)^2\right), \quad (8.8)$$

The proof is based on a relatively standard analysis of the convergence of the stochastic gradient method by Nemirovski et al. [Nem+09] (see also [WR22; RSS12] for expositions of the proof technique).

In a nutshell, from the noise2noise self-supervised loss we can compute stochastic gradients of the risk as $\nabla \ell_{\mathrm{SS}}(f_\theta(\mathbf{y}_i), \mathbf{y}_i')$, where $(\mathbf{y}_i, \mathbf{y}_i')$ is a training example. This gradient is an unbiased estimate of the gradient of the risk $\nabla R(\boldsymbol{\theta})$. The variance of the stochastic gradient determines the rate on the right-hand side of Equation (8.8).

Theorem 9 establishes that the risk is upper bounded by a problem-dependent noise floor, which is the risk of the optimal estimator, $R(\mathbf{W}^*)$, plus a term associated with minimizing a self-supervised loss constructed from finitely many training examples. This term decays as c/N, which is exactly the rate we see in simulations for this setup. Moreover, the term associated with minimizing the self-supervised loss becomes larger in the noise variance σ_e^2, and thus reflects that the self-supervised loss is a worse approximation of the supervised loss, as σ_e^2 increases.

8.2.1 Concluding remarks

Designing a self-supervised loss that approximates the risk (8.2) is a principled approach to learning a neural network for signal reconstruction, and works very well, given sufficient training data. A gradient computed on a self-supervised loss (i.e., $\nabla \ell_{\mathrm{SS}}(f_\theta(\mathbf{y}_i), \mathbf{y}'_i)$) is in general a noisier approximation than a gradient computed on a supervised loss (i.e., $\nabla \ell(f_\theta(\mathbf{y}_i), \mathbf{x}_i)$) and thus more training examples are required when training with a self-supervised loss to achieve the same performance. The better the approximation, the fewer training examples are required. For example, for the denoising example, if the noise vector $\mathbf{e}'$ has smaller variance, fewer training examples are required for the same performance. Similarly, for the compressive sensing example, if the number of non-zero frequencies associated with the estimate $\mathbf{y}'$ is larger, then fewer training examples are required to train a network with the same performance.

8.3 Stein's unbiased estimator

Stein's unbiased risk estimator [Ste81] (SURE) provides an unbiased estimate of the mean-squared error for an estimator of the mean of a Gaussian random vector. The estimator can be used to formulate a self-supervised loss for training. Here, we discuss how a self-supervised loss for denoising and compressive sensing can be obtained based on Stein's unbiased estimator.

8.3.1 Denoising

Suppose our goal is to estimate a vector $\mathbf{x} \in \mathbb{R}^n$ from a noisy observation $\mathbf{y} = \mathbf{x} + \mathbf{e}$, where $\mathbf{e} \sim \mathcal{N}(0, \sigma^2 \mathbf{I})$. Consider an estimator $f_\theta \colon \mathbb{R}^n \to \mathbb{R}^n$ parameterized by $\boldsymbol{\theta}$ for estimating the signal $\mathbf{x}$ from the noisy observation $\mathbf{y}$. The expectation of the mean-squared error with respect to the random noise can be written as [Ste81]

$$\mathbb{E}_{\mathbf{e}}\left[\|f_\theta(\mathbf{y}) - \mathbf{x}\|_2^2\right] = \mathbb{E}_{\mathbf{e}}\left[\|f_\theta(\mathbf{y}) - \mathbf{y}\|_2^2\right] + 2\sigma^2 \mathrm{div}_{\mathbf{y}}(f_\theta) - n\sigma^2, \tag{8.9}$$

where the divergence of the function f_θ is defined as

$$\mathrm{div}_{\mathbf{y}}(f_\theta) = \sum_{i=1}^{n} \frac{\partial [f_\theta(\mathbf{y})]_i}{\partial y_i}. \tag{8.9}$$

Note that the first term in the SURE-loss (8.9) is small if the estimate predicts the observation well, and the second term penalizes the denoiser for varying too much as a function of the input.

The right-hand side of the SURE-loss (8.9) is only a function of the measurement $\mathbf{y}$ and does not depend on the signal $\mathbf{x}$. Thus, this formulation suggests that it is possible to train a Gaussian denoiser only from noisy examples by minimizing the right-hand side of the SURE-loss (8.9). Specifically, given N noisy observations, we can minimize the self-supervised loss

$$\mathcal{L}_{\text{SURE}}(\boldsymbol{\theta}) = \sum_{i=1}^{N} \left(\| f_{\boldsymbol{\theta}}(\mathbf{y}_i) - \mathbf{y}_i \|_2^2 + 2\sigma^2 \text{div}_{\mathbf{y}_i}(f_{\boldsymbol{\theta}}) - n\sigma^2 \right).$$

This loss works well for training a Gaussian denoiser. Specifically, a neural network trained on sufficiently many training examples for Gaussian denoising performs essentially equally well as the same network trained on the standard supervised training objective, which is $\frac{1}{N}\sum_{i=1}^{N} \| f_{\boldsymbol{\theta}}(\mathbf{y}_i) - \mathbf{x}_i \|_2^2$ [Met+20]. However, this self-supervised loss relies on the noise being Gaussian (or more generally, coming from an exponential family), and the loss does not generalize in a straightforward way to non-Gaussian distributions. Thus, while compared to the noise2noise self-supervised loss discussed in Section 8.1.1, the SURE-self-supervised loss only requires a measurement (and not a pair of measurements); the loss makes stronger assumptions (i.e., that the noise is Gaussian with known variance).

8.3.2 Compressive sensing

Eldar [Eld09] proposed a generalized SURE-loss which allows to formulate a generalized SURE-loss that applies to a setup where we are given noisy linear measurements of a signal $\mathbf{y} = \mathbf{A}\mathbf{x} + \mathbf{e}$, where $\mathbf{e} \sim \mathcal{N}(0, \sigma^2\mathbf{I})$ is Gaussian noise with known variance σ^2. This generalized SURE-loss allows to train a neural network for compressive sensing, without any ground-truth images. For simulated data, this performs similar to training on a supervised loss [Met+20]; as for the denoising case, this relies on the noise being Gaussian.

8.4 Masking-based approaches

Another set of self-supervised methods is based on constructing a self-supervised loss that predicts certain parts of an image based on other parts of an image. Such approaches are conceptually different from the noise2noise like and SURE-based approaches discussed so far in that they explicitly or implicitly make assumptions about the images itself and about the noise (for example, that the noise is pixel-wise independent). Those masking-based methods also typically pertain to denoising.

Noise2self [BR19] and Noise2void [KBJ19] are examples of such an approach for training a network in an unsupervised fashion. For both methods, the idea is to train a network f_θ for denoising by minimizing the self-supervised loss

$$\mathcal{L}_{\mathrm{SS}}(\boldsymbol{\theta}) = \frac{1}{N}\sum_{i=1}^{N} \ell(f_\theta(\mathbf{M}_i\mathbf{y}_i), \mathbf{y}_i),$$

where $\mathbf{y}_i = \mathbf{x}_i + \mathbf{e}_i$ are noisy training examples, and $\mathbf{M}_i$ is a random mask that masks some pixels, and ℓ is a loss function such as the mean-squared error, for example. This approach is often called blind-spot prediction, since the network is trained to predict blind spots in the image based on neighbouring pixels. Such masking-based approaches work relatively well for Gaussian denoising, for example, but do not work as well for setups where the noise is dependent across pixels.

Another recent approach, called neighbor2neighbor [Hua+21], is based on training a network to predict one down-sampled image from another down-sampled image. Let $(D_1(\mathbf{y}), D_2(\mathbf{y})) \in \mathbb{R}^{n/2 \times n/2}$ be a pair of images generated from a single noisy image $\mathbf{y} \in \mathbb{R}^{n \times n}$. The operators D_1 and D_2 perform down-sampling, for example by taking two randomly chosen pixels from each 2×2 patch and assigning one of the pixels to $D_1(\mathbf{y})$ and the other to $D_2(\mathbf{y})$. Another example is to simply average the diagonal pixel and assign one average to $D_1(\mathbf{y})$ and the other to $D_2(\mathbf{y})$. A convolutional neural network is then trained on the self-supervised loss to predict one image based on the other, by minimizing the self-supervised loss

$$\mathcal{L}_{\mathrm{SS}}(\boldsymbol{\theta}) = \frac{1}{N}\sum_{i=1}^{N} \ell(f_\theta(D_1(\mathbf{y}_i)), D_2(\mathbf{y}_i)).$$

After training, the network is applied to estimate a clean image from a noisy observation $\mathbf{y}$ according to $\hat{\mathbf{x}} = f_\theta(\mathbf{y})$.

9
Signal reconstruction by imposing generative priors

In the previous three chapters, we discussed neural networks trained in a supervised fashion to perform signal reconstruction. The resulting networks take a measurement as an input and yield the estimate of the signal as an output. Those networks perform very well for signal reconstruction tasks, but typically pertain to a single task or forward operator.

For example, if a network is trained in a supervised fashion to reconstruct an image from 4x undersampled (accelerated) MRI data, then the network performs relatively poorly on 3x undersampled MRI data. As a second example, if a network is trained for denoising on data collected with a given signal-to-noise ratio, it typically performs relatively poorly on data with a different signal-to-noise ratio.

In this chapter, we discuss an alternative approach to signal reconstruction based on learning a (generative) signal prior and imposing it during signal reconstruction. This approach is more versatile than a neural network trained end-to-end for signal reconstruction, but requires (often significantly) more computational resources at inference. Moreover, in controlled benchmark settings like the fastMRI competition, approaches based on generative priors also currently perform slightly worse than neural networks trained end-to-end, although that can depend on the setup of the benchmark.

Recall that in order to reconstruct a signal $\mathbf{x}$ from an undersampled or noisy measurement $\mathbf{y}$, we make explicit or implicit assumptions on the structure of the signal $\mathbf{x}$. A popular explicit assumption discussed in Chapter 4 is that the signal is sparse in some basis. However, assuming a sparse model often induces a modelling error as, for example, images are only approximately sparse in a given basis or dictionary. In order to improve performance, a natural idea is to learn a signal model, and impose this signal model during reconstruction.

The general idea is as follows. In the training phase, we learn a signal model in the form of a neural network based on a set of example signals. The examples can be a set of natural images for an imaging problem, or a set of clean MRI images for accelerated MRI reconstruction. This signal model is a neural network that often maps a latent vector, $\mathbf{z} \in \mathbb{R}^k$, to a signal of interest, for example an image. The latent vector is a representation of the image and is either low-dimensional, or has other simple structure. To reconstruct an image given a noisy measurement $\mathbf{y} = \mathbf{A}\mathbf{x} + \mathbf{e}$, we find a reconstructed signal $\hat{\mathbf{x}}$ that is likely according to the signal

Deep Learning for Computational Imaging. Reinhard Heckel, Oxford University Press. © Reinhard Heckel (2025).
DOI: 10.1093/oso/9780198947172.003.0009

model, and at the same time is consistent (or approximately consistent) with the measurement, i.e., $\mathbf{A}\hat{\mathbf{x}} \approx \mathbf{y}$.

In the remainder of this chapter, we first briefly discuss autoencoders as a simple signal model, and then give a brief overview of generative signal models. We then discuss how signal reconstruction can be performed by imposing a signal model like an autoencoder, and discuss the general idea of signal reconstruction with generative models. Finally, we discuss a variety of generative models in more detail, and how to use them to perform image reconstruction.

9.1 Autoencoders

Perhaps the simplest approach to learn an image model parameterized by a neural network is to train an autoencoder consisting of an encoder and a decoder. The encoder $E_{\boldsymbol{\vartheta}}: \mathbb{R}^n \to \mathbb{R}^k$ is a neural network that maps the signal to a latent vector, and the decoder $G_{\boldsymbol{\theta}}: \mathbb{R}^k \to \mathbb{R}^n$ maps the latent vector back to the signal. Let $\boldsymbol{\vartheta}$ and $\boldsymbol{\theta}$ be the parameters of the encoder and decoder, respectively. The encoder and decoder are trained on the set of example images $\{\mathbf{x}_1, \ldots, \mathbf{x}_N\}$ by minimizing the loss

$$\mathcal{L}(\boldsymbol{\vartheta}, \boldsymbol{\theta}) = \frac{1}{n} \sum_{i=1}^{n} \|\mathbf{x}_i - G_{\boldsymbol{\theta}}(E_{\boldsymbol{\vartheta}}(\mathbf{x}_i))\|_2^2$$

with gradient descent or another iterative optimization algorithm. This yields an encoder $E_{\hat{\boldsymbol{\vartheta}}}$ and a decoder $G_{\hat{\boldsymbol{\theta}}}$, where the notation $\hat{\boldsymbol{\vartheta}}, \hat{\boldsymbol{\theta}}$ indicates that the parameters are learned based on the training data.

The decoder is the learned image model and enables us to represent images with a low-dimensional latent vector. This image model can be better than a hand-crafted image model like a sparse representation, through learning from data. The image model can be used for image reconstruction as discussed below. The image model $G_{\hat{\boldsymbol{\theta}}}$ learned with an autoencoder does not, however, enable us to sample likely images or generate entirely new images that resemble images from the dataset and thus is not considered a generative model.

9.2 Overview of generative signal models

Generative models parameterize a probabilistic data generation process. A generative model trained on a dataset enables us to draw samples that resemble the examples from the dataset. For example, a generative model trained on images of faces enables us to generate new images of faces. The faces generated by modern generative models are essentially indistinguishable from real faces.

Below, we briefly summarize the currently most prevalent generative models for image generation, and how to sample from them. In the remainder of this chapter and in the next chapter, we discuss those generative models in more detail, and explore their application in image reconstruction.

- **Variational autoencoders (VAEs).** Variational autoencoders [KW19] are a class of generative models that are trained on a set of examples $\mathbf{x}_1, \ldots, \mathbf{x}_n \in \mathbb{R}^n$ and yield a decoder $G: \mathbb{R}^k \to \mathbb{R}^n$. With the decoder, we can generate new examples as $G(\mathbf{z})$, where $\mathbf{z}$ is a latent vector typically drawn from a Gaussian distribution with zero-mean and identity co-variance matrix. The decoder and a stochastic encoder (both parameterized by a neural network), are jointly trained. We discuss VAEs in detail in Section 9.5.
- **Generative adversarial networks (GANs).** Generative adversarial networks [Goo+14] are a class of generative models trained in a game-theoretic adversarial way. GANs consist of a discriminator and a generator $G: \mathbb{R}^k \to \mathbb{R}^n$, both are parameterized by neural networks. The generator maps a k-dimensional latent space to an n-dimensional signal space. Once trained, $G(\mathbf{z})$ is a sample from the learned data distribution, where the latent vector $\mathbf{z}$ is from a Gaussian distribution $\mathcal{N}(0, \mathbf{I})$. See Section 9.6 for details.
- **Flow-based generative models.** Flow-based generative models or flow-based invertible neural networks [DKB15, DSB17, KD18] are a class of generative models that are parameterized by a neural network $G: \mathbb{R}^n \to \mathbb{R}^n$ that is invertible or one-to-one by design, i.e., for each output there is a unique input. Once trained, a sample of the model is $G(\mathbf{z})$, where $\mathbf{z}$ is a sample from the latent distribution. Flow-based models provide a direct estimate of the probability density, as each output corresponds uniquely to an input, which enables precise density estimation.
- **Diffusion and score-based generative models.** Diffusion models [Soh+15] are probabilistic models that parameterize a data distribution $p(\mathbf{x})$ by a model that gradually denoises a random variable that is typically Gaussian. Diffusion models consist of a forward diffusion process that gradually adds noise to the input and a reverse denoising process that is aimed at inverting the diffusion process. We discuss diffusion models in Chapter 10.

For imaging applications, the generative networks (VAEs, GANs, Flow-based, and diffusion models) are usually parameterized with a convolutional network, but parameterizations with transformers or networks that combine building blocks from convolutional networks and from transformers (such as self-attention) also work well.

9.3 Signal reconstruction by imposing a signal model

In this section, we discuss how a signal model parameterized by a neural network can be used to reconstruct an image from a noisy measurement $\mathbf{y} = \mathbf{A}\mathbf{x} + \mathbf{e}$. We discuss reconstruction with a signal model that does not necessarily have a probabilistic interpretation, such as the decoder from an autoencoder. In the next section, we discuss reconstruction with generative models that represent probability distributions.

Assume that the signal $\mathbf{x}$ to be reconstructed is in the range of a signal model $G: \mathbb{R}^d \to \mathbb{R}^n$, i.e., there is a latent vector $\mathbf{z} \in \mathbb{R}^d$ so that $\mathbf{x} = G(\mathbf{z})$. For concreteness, suppose that the model was trained in an autoencoder fashion. For now, we do not assume that the signal model is a generative model, i.e., we do not assume that the signal model parameterizes a probability distribution over images.

With the assumption that the signal to be reconstructed lies in the range of the signal model, we reconstruct a signal from the noisy measurement by estimating the latent variable by minimizing the loss

$$\mathcal{L}(\mathbf{z}) = \frac{1}{2}\|\mathbf{A}G(\mathbf{z}) - \mathbf{y}\|_2^2.$$

In practice, we can minimize the loss with gradient descent or another first-order optimizer such as Adam. From this minimization, we obtain an estimate $\hat{\mathbf{z}}$ of the latent variable, and with this we estimate the signal as $\hat{\mathbf{x}} = G(\hat{\mathbf{z}})$.

In order to understand how well this approach works, it is illustrative to consider a simple signal model in the form of a linear subspace $G(\mathbf{z}) = \mathbf{U}\mathbf{z}$, and study reconstruction of an arbitrary signal $\mathbf{x}$, typically lying close to the subspace, from a noisy linear measurement $\mathbf{y} = \mathbf{A}\mathbf{x} + \mathbf{e} \in \mathbb{R}^m$. Here, $\mathbf{U} \in \mathbb{R}^{n \times d}$ is an orthonormal basis for the subspace in $\mathbb{R}^n$. Typically, the subspace is low-dimensional, i.e., $d < n$.

Suppose that the matrix $\mathbf{AU}$ has full column rank. Note that this implies that the number of measurements cannot be smaller than the subspace dimension, i.e., $m \geq d$. Under this assumption, the unique minimizer of the loss $\mathcal{L}(\mathbf{z})$ is $\hat{\mathbf{z}} = (\mathbf{AU})^\dagger \mathbf{y}$, where $(\cdot)^\dagger$ is the pseudo-inverse, and with this the signal estimate is $\hat{\mathbf{x}} = \mathbf{U}(\mathbf{AU})^\dagger \mathbf{y}$. A bound on the mean-square error of the estimate $\hat{\mathbf{x}}$ is

$$\|\hat{\mathbf{x}} - \mathbf{x}\|_2 \leq (1 + \xi) \min_{\mathbf{z}} \|G(\mathbf{z}) - \mathbf{x}\|_2 + \|(\mathbf{AU})^\dagger \mathbf{e}\|_2, \tag{9.1}$$

where we defined $\xi = \|(\mathbf{AU})^\dagger \mathbf{A}\|$ for notational convenience. Recall that $\|\mathbf{B}\|$ is the matrix or operator norm which is the largest singular value of the matrix $\mathbf{B}$. The proof of the bound (9.1) is at the end of this section.

If $\mathbf{A} \in \mathbb{R}^{m \times n}$ is a Gaussian random matrix with $m > cd$, where c is a numerical constant, then $(\mathbf{AU})^\dagger$ is well-conditioned and thus ξ is close to one. Note that the condition $m > cd$ simply requires the number of measurements to be larger by a constant factor than the dimension of the subspace, which is a natural requirement.

The bound (9.1) can be interpreted as follows: The first term is a representation error that describes the error due to assuming (for signal reconstruction) that the signal $\mathbf{x}$ lies in the range of the signal model, i.e., lies in the subspace for our example. The second term is an error term due to the noise. The representation error persists irrespective of how many measurements we take. This illustrates that signal reconstruction with an image model reaches an error floor, where the reconstruction quality does not improve in the number of measurements if the signal to be reconstructed does not lie in the range of the prior. On the contrary, for ℓ_1-regularized least-squares the performance improves in the number of measurements.

In practice, G is a neural network and thus a non-linear map. However, the intuition based on the subspace model applies as follows: In the regime where the number of measurements is small, imposing a trained prior often yields better image reconstruction performance than using ℓ_1-minimization, but as the number of measurements increases, the performance plateaus when the representation error dominates. In contrast, the performance of ℓ_1-regularized least-squares increases in the number of measurements and eventually surpasses that of imposing a generative prior. That is because the representation error vanishes for a large number of measurements, where the optimal regularization parameter of the ℓ_1-regularizer goes to zero.

Proof of Equation (9.1): *For* $\hat{\mathbf{x}} = \mathbf{U}(\mathbf{A}\mathbf{U})^\dagger \mathbf{y}$ *and* $\mathbf{y} = \mathbf{A}\mathbf{x} + \mathbf{e}$, *we have*

$$\begin{aligned}
\|\hat{\mathbf{x}} - \mathbf{x}\|_2 &= \|\mathbf{U}(\mathbf{A}\mathbf{U})^\dagger \mathbf{y} - \mathbf{x}\|_2 \\
&= \|\mathbf{U}(\mathbf{A}\mathbf{U})^\dagger (\mathbf{A}(\mathbf{U}\mathbf{U}^T\mathbf{x} + (\mathbf{I} - \mathbf{U}\mathbf{U}^T)\mathbf{x}) + \mathbf{e}) - \mathbf{x}\|_2 \\
&= \|(\mathbf{U}\mathbf{U}^T - \mathbf{I})\mathbf{x} + \mathbf{U}(\mathbf{A}\mathbf{U})^\dagger \left(\mathbf{A}(\mathbf{I} - \mathbf{U}\mathbf{U}^T)\mathbf{x} + \mathbf{e}\right) \|_2 \\
&\leq (1 + \xi) \|(\mathbf{I} - \mathbf{U}\mathbf{U}^T)\mathbf{x}\|_2 + \|(\mathbf{A}\mathbf{U})^\dagger \mathbf{e}\|_2 \\
&= (1 + \xi) \min_{\mathbf{z}} \|G(\mathbf{z}) - \mathbf{x}\|_2 + \|(\mathbf{A}\mathbf{U})^\dagger \mathbf{e}\|_2,
\end{aligned}$$

where we used $\xi = \|(\mathbf{A}\mathbf{U})^\dagger \mathbf{A}\|_2$. *This concludes the proof. For the last equation, we used that*

$$\begin{aligned}
\min_{\mathbf{z}} \|G(\mathbf{z}) - \mathbf{x}\|_2 &= \min_{\mathbf{z}} \|\mathbf{U}\mathbf{z} - \mathbf{x}\|_2 \\
&= \|\mathbf{U}\mathbf{U}^T\mathbf{x} - \mathbf{x}\|_2 \\
&= \|(\mathbf{U}\mathbf{U}^T - \mathbf{I})\mathbf{x}\|_2.
\end{aligned}$$

9.4 Signal reconstruction with a generative model

Now suppose we have access to a pre-trained generative network G that maps a latent variable to an image. Generative networks parameterize probability distributions over a class of images. Some pre-trained networks, such as those from VAEs and flow-based invertible networks, explicitly provide access to the probability distribution, and others implicitly provide access by enabling sampling, such as GANs.

If the generative network gives explicit access to the probability distribution p, a natural reconstruction approach is to perform maximum likelihood estimation by solving the following optimization problem (see Section 2.1)

$$\hat{\mathbf{x}} = \arg\min_{\mathbf{x}} \frac{1}{2}\|\mathbf{A}\mathbf{x} - \mathbf{y}\|_2^2 - \lambda \log p(\mathbf{x}), \tag{9.2}$$

where p is the likelihood function over the signal $\mathbf{x}$, which in our case is given through the generative model, and λ is a regularization parameter.

However, it can be challenging to optimize over the parameter $\mathbf{x}$, and GANs, for example, do not provide explicit access to the likelihood function $\log p(\mathbf{x})$, because they do not model the likelihood explicitly, as we will see later. Thus the majority of approaches for signal reconstruction with GANs and flow-based invertible networks solve an optimization problem over the latent space that promotes a high likelihood via a penalty on the latent variable, often by minimizing

$$\hat{\mathbf{z}} = \arg\min_{\mathbf{z}} \frac{1}{2}\|\mathbf{A}G(\mathbf{z}) - \mathbf{y}\|_2^2 + \lambda\|\mathbf{z}\|_2^2. \tag{9.3}$$

From the estimate for the latent variable, $\hat{\mathbf{z}}$, the estimate of the signal can be obtained as $\hat{\mathbf{x}} = G(\hat{\mathbf{z}})$.

This approach has been proposed with a GAN generator by Bora et al. [Bor+17] and with an invertible flow-based generator by Asim et al. [Asi+20]. Menon et al. [Men+20] proposed to use a GAN generator with a slightly different penalty that ensures that the latent vector lies on or close to a unit sphere.

Generative networks perform excellent for imaging problems where the goal is to obtain a realistic image, even if the reconstructed image is not entirely accurate. For example, suppose we want perform image super-resolution, i.e., obtain a high-resolution image from a low-resolution image. In a consumer photography application, we might not primarily be interested in accuracy, but in obtaining a realistic image that is consistent with the low-resolution image. For this application, signal reconstruction with a generative prior yields very good looking reconstructions.

In medical imaging, however, our primary goal is to obtain an accurate image. For example, if we image the knee of a patient, we want to be as accurate as possible, i.e., obtain an image that is as close as possible to the actual knee that we are imaging, and we are not primarily interested in obtaining a realistic looking knee that is consistent with the measurements. For such applications, in particular MRI, signal reconstruction with generative priors currently performs better than total-variation minimization, but worse than training a neural network end-to-end for reconstruction.

9.5 Variational autoencoders

Variational autoencoders are a type of generative model introduced by Kingma and Welling [KW14] and Rezende et al. [RMW14]. Kingma and Welling [KW19] provide an extensive introduction.

Given a dataset of signals or images $\{\mathbf{x}_1, \ldots, \mathbf{x}_N\}$ variational autoencoders model the distribution of the signals or images via a probabilistic latent variable model that is learned from the data. VAEs consist of latent variables that follow a simple distribution (often, they are zero-mean and unit-variance Gaussian), and a model $G_\theta \colon \mathbb{R}^k \to \mathbb{R}^n$ with parameters $\boldsymbol{\theta}$ (called generator or decoder) that maps the latent parameter to a signal. The generator is typically a neural network. VAEs can model complex distributions of data, such as natural images, and can be used to sample new examples from the distribution, or to impose a prior to reconstruct a signal or image from examples, as described above. In this section, we give a brief introduction to VAEs.

We assume a noisy observational model, where the distribution of the signal given the latent variable, $p_\theta(\mathbf{x}|\mathbf{z})$, is Gaussian with mean $G_\theta(\mathbf{z})$ and diagonal co-variance matrix $\sigma^2\mathbf{I}$. We assume a simple prior $p(\mathbf{z})$ on the latent variable $\mathbf{z}$. Specifically we assume that the latent variable consists of independent $\mathcal{N}(0,1)$ variables, which induces a joint distribution $p_\theta(\mathbf{x}, \mathbf{z}) = p_\theta(\mathbf{x}|\mathbf{z})p(\mathbf{z})$. This model presumes that the data is generated by first drawing a Gaussian latent variable $\mathbf{z} \sim p(\mathbf{z})$, second computing the mean $G_\theta(\mathbf{z})$, and third drawing the signal $\mathbf{x}$ from the Gaussian distribution $\mathcal{N}(G_\theta(\mathbf{z}), \sigma^2\mathbf{I})$. The last step is omitted when sampling from the distribution, but is used to justify the training loss.

We next discuss how to fit such a latent variable model. A natural approach to fit a latent variable model would be to maximize the likelihood of the observed data, for example by maximizing the log-likelihood of the observed data

$$\arg\max_{\theta} \sum_{i=1}^{N} \log p_\theta(\mathbf{x}_i)$$

with gradient ascent. However, optimizing the likelihood or the log-likelihood directly is computationally intractable since we have no efficient way of computing gradients of the likelihood or log-likelihood.

Variational encoders use a stochastic encoder $q_\vartheta(\mathbf{z}|\mathbf{x})$ which is typically parameterized by a neural network E_ϑ with parameters $\boldsymbol{\vartheta}$. The encoder is stochastic in that it generates a latent random variable and not a deterministic variable, and consists

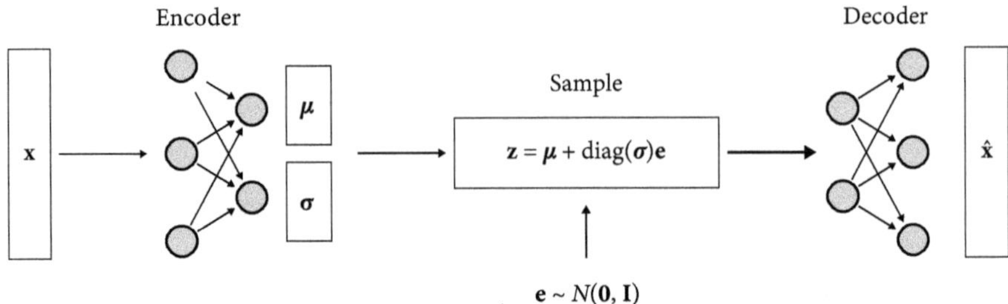

Figure 9.1 A variational autoencoder enables training a latent variable model by training encoder and decoder networks so that the encoder maps the signal to mean and variance parameters, which are used to sample the latent variable $\mathbf{z}$ from a Gaussian with mean $\boldsymbol{\mu}$ and co-variance matrix diag($\boldsymbol{\sigma}$). This latent variable $\mathbf{z}$ is then mapped to an image $\hat{\mathbf{x}}$ with the decoder. After training, the decoder network fed with a standard normal Gaussian latent vector $\mathbf{z}$ yields samples from the distribution.

of an encoder network $E_\vartheta \colon \mathbb{R}^n \to \mathbb{R}^{2k}$ which maps a signal $\mathbf{x}$ to a mean parameter $\boldsymbol{\mu}_\vartheta(\mathbf{x})$ and a variance parameter $\boldsymbol{\sigma}_\vartheta(\mathbf{x})$.

To sample from the distribution $q_\vartheta(\mathbf{z}|\mathbf{x})$, i.e., to generate a latent variable $\mathbf{z}$ corresponding to the signal $\mathbf{x}$, we first evaluate the encoder network, which yields the mean $\boldsymbol{\mu}_\vartheta(\mathbf{x})$ and variance $\boldsymbol{\sigma}_\vartheta(\mathbf{x})$, and then sample from a Gaussian with mean $\boldsymbol{\mu}_\vartheta(\mathbf{x})$ and diagonal co-variance matrix diag($\boldsymbol{\sigma}_\vartheta(\mathbf{x})$). See Figure 9.1 for an illustration of a variational autoencoder.

The parameters of the encoder are optimized so that the distribution $q_\vartheta(\mathbf{z}|\mathbf{x})$ approximates the distribution $p_\theta(\mathbf{z}|\mathbf{x})$. The encoder network and associated distribution $q_\vartheta(\mathbf{z}|\mathbf{x})$ are introduced because it is computationally intractable to compute and work with the distribution $p_\theta(\mathbf{z}|\mathbf{x})$ directly.

9.5.1 The evidence lower bound (ELBO)

The encoder and decoder networks are trained by maximizing a proxy for the likelihood called the evidence lower bound (ELBO). The log-likelihood for a given data point $\mathbf{x}$ is given by [KW19]

$$\begin{aligned}
\log p_\theta(\mathbf{x}) &\overset{\text{(i)}}{=} \mathbb{E}_{\mathbf{z}\sim q_\vartheta(\mathbf{z}|\mathbf{x})}\left[\log p_\theta(\mathbf{x})\right] \\
&\overset{\text{(ii)}}{=} \mathbb{E}_{\mathbf{z}\sim q_\vartheta(\mathbf{z}|\mathbf{x})}\left[\log \frac{p_\theta(\mathbf{x},\mathbf{z})}{p_\theta(\mathbf{z}|\mathbf{x})}\right] \\
&= \mathbb{E}_{\mathbf{z}\sim q_\vartheta(\mathbf{z}|\mathbf{x})}\left[\log \frac{p_\theta(\mathbf{x},\mathbf{z})}{q_\vartheta(\mathbf{z}|\mathbf{x})}\frac{q_\vartheta(\mathbf{z}|\mathbf{x})}{p_\theta(\mathbf{z}|\mathbf{x})}\right] \\
&= \underbrace{\mathbb{E}_{\mathbf{z}\sim q_\vartheta(\mathbf{z}|\mathbf{x})}\left[\log \frac{p_\theta(\mathbf{x},\mathbf{z})}{q_\vartheta(\mathbf{z}|\mathbf{x})}\right]}_{\mathrm{ELBO}_{\vartheta,\theta}(\mathbf{x})} + \underbrace{\mathbb{E}_{\mathbf{z}\sim q_\vartheta(\mathbf{z}|\mathbf{x})}\left[\frac{q_\vartheta(\mathbf{z}|\mathbf{x})}{p_\theta(\mathbf{z}|\mathbf{x})}\right]}_{D_{\mathrm{KL}}(q_\vartheta(\mathbf{z}|\mathbf{x})\|p_\theta(\mathbf{z}|\mathbf{x}))},
\end{aligned} \tag{9.4}$$

where equation (i) holds since the log-likelihood of $p_\theta(\mathbf{x})$ does not depend on $\mathbf{z}$, so we can take expectation over $\mathbf{z}$, and equation (ii) holds by the definition of the conditional density.

The second term in Equation (9.4) is the Kullback-Leibler (KL) divergence between the distributions $q_\vartheta(\mathbf{z}|\mathbf{x})$ and $p_\theta(\mathbf{z}|\mathbf{x})$. The KL divergence

$$D_{\mathrm{KL}}(q \parallel p) = \int q(x) \log \frac{q(x)}{p(x)} dx \tag{9.5}$$

is a measure of how far the distribution q is from the distribution p and is equal to zero if and only if $q = p$.

The KL divergence is non-negative, i.e., $D_{\mathrm{KL}}(q \parallel p) \geq 0$, and it does not commute, i.e., in general $D_{\mathrm{KL}}(q \parallel p) \neq D_{\mathrm{KL}}(p \parallel q)$. In our setup, the KL divergence is zero if and only if the distribution $q_\vartheta(\mathbf{z}|\mathbf{x})$ equals the true posterior distribution $p_\theta(\mathbf{z}|\mathbf{x})$.

The first term in Equation (9.4) is called the variational lower bound or evidence lower bound (ELBO) and is a lower bound on the log-likelihood of the data.

9.5.2 Interpretation of ELBO

Rearranging Equation (9.4) gives

$$\mathrm{ELBO}_{\vartheta,\theta}(\mathbf{x}) = \log p_\theta(\mathbf{x}) - D_{\mathrm{KL}}(q_\vartheta(\mathbf{z}|\mathbf{x}), p_\theta(\mathbf{z}|\mathbf{x})). \tag{9.6}$$

Thus, maximizing the ELBO approximately maximizes the log-likelihood $\log p_\theta(\mathbf{x})$, thus, the generative model improves, and also approximately minimizes the KL divergence between the approximate $q_\vartheta(\mathbf{z}|\mathbf{x})$ and the true posterior $p_\theta(\mathbf{z}|\mathbf{x})$, so the approximate posterior improves. Variational autoencoders maximize the ELBO instead of the log-likelihood of the data directly.

Let us take a closer look at the ELBO. The ELBO can be written as

$$\begin{aligned} \mathrm{ELBO}_{\vartheta,\theta}(\mathbf{x}) &= \mathbb{E}_{\mathbf{z}\sim q_\vartheta(\mathbf{z}|\mathbf{x})}\left[\log \frac{p_\theta(\mathbf{x}|\mathbf{z})p(\mathbf{z})}{q_\vartheta(\mathbf{z}|\mathbf{x})}\right] \\ &= \mathbb{E}_{\mathbf{z}\sim q_\vartheta(\mathbf{z}|\mathbf{x})}\left[\log p_\theta(\mathbf{x}|\mathbf{z})\right] + \mathbb{E}_{\mathbf{z}\sim q_\vartheta(\mathbf{z}|\mathbf{x})}\left[\log \frac{p(\mathbf{z})}{q_\vartheta(\mathbf{z}|\mathbf{x})}\right], \end{aligned} \tag{9.7}$$

where the first term can be interpreted as a reconstruction error term and the second term as a regularizer. To see this, recall that our observation model assumes that $p_\theta(\mathbf{x}|\mathbf{z})$ is a Gaussian distribution with mean $G_\theta(\mathbf{z})$ and co-variance matrix $\sigma^2\mathbf{I}$. Thus,

$$\log p_\theta(\mathbf{x}|\mathbf{z}) = -\frac{1}{2\sigma^2}\|\mathbf{x} - G_\theta(\mathbf{z})\|_2^2 + \text{const}, \tag{9.8}$$

and therefore the expectation $\mathbb{E}_{q_\vartheta(\mathbf{z}|\mathbf{x})}[\log p_\theta(\mathbf{x}|\mathbf{z})]$ can be interpreted as the expected ℓ_2-reconstruction error under the encoder model.

The second term in Equation (9.7) is

$$\begin{aligned}\mathbb{E}_{q_\vartheta(\mathbf{z}|\mathbf{x})}\left[\log \frac{p(\mathbf{z})}{q_\vartheta(\mathbf{z}|\mathbf{x})}\right] &= -D_{KL}(q_\vartheta(\mathbf{z}|\mathbf{x}), p(\mathbf{z})) \\ &= \sum_{i=1}^{k} \frac{1}{2}\left(1 + \log \sigma_{\vartheta,i}^2(\mathbf{x}) - \mu_{\vartheta,i}^2(\mathbf{x}) - \sigma_{\vartheta,i}^2(\mathbf{x})\right),\end{aligned}$$

where we used that z_i is $\mathcal{N}(0,1)$ distributed, and subsequently we used the formula for the KL divergence between two Gaussians. Here, $\mu_{\vartheta,i}(\mathbf{x})$ and $\sigma_{\vartheta,i}(\mathbf{x})$ are the outputs of the encoder with index i. This term can be interpreted as a regularization term which encourages the two distributions to be similar.

Thus, Equation (9.7) can be written as

$$\begin{aligned}\mathrm{ELBO}_{\vartheta,\theta}(\mathbf{x}) &= \mathbb{E}_{\mathbf{z}\sim q_\vartheta(\mathbf{z}|\mathbf{x})}[\log p_\theta(\mathbf{x}|\mathbf{z})] \\ &\quad + \sum_{i=1}^{k} \frac{1}{2}\left(1 + \log \sigma_{\vartheta,i}^2(\mathbf{x}) - \mu_{\vartheta,i}^2(\mathbf{x}) - \sigma_{\vartheta,i}^2(\mathbf{x})\right). \end{aligned} \tag{9.9}$$

9.5.3 Optimizing the ELBO with stochastic gradient descent

In order to fit a model, we maximize the ELBO over our training dataset, i.e., we solve the optimization problem

$$\arg\max_{\theta,\vartheta} \sum_{i=1}^{n} \mathrm{ELBO}_{\vartheta,\theta}(\mathbf{x}_i). \tag{9.10}$$

In principle, we could apply gradient ascent to maximize the objective, but computing the gradients is infeasible as it would involve computing expectations. However, it is cheap to compute stochastic gradients, i.e., random variables that in expectation are equal to the gradient of the ELBO. We therefore work with those stochastic gradients, i.e., apply the stochastic gradient method.

In order to compute a stochastic gradient with respect to the parameter $\boldsymbol{\theta}$ of the ELBO expression in Equation (9.9), we note that

$$\begin{aligned}\nabla_\theta \mathrm{ELBO}_{\vartheta,\theta}(\mathbf{x}) &= \nabla_\theta \mathbb{E}_{q_\vartheta(\mathbf{z}|\mathbf{x})}[\log p_\theta(\mathbf{x}|\mathbf{z})] \\ &= \mathbb{E}_{q_\vartheta(\mathbf{z}|\mathbf{x})}[\nabla_\theta \log p_\theta(\mathbf{x}|\mathbf{z})].\end{aligned}$$

Computing this gradient is intractable due to the expectation. However, it is cheap to compute a stochastic gradient for a given $\mathbf{x}$ by sampling $\mathbf{z} \sim q_\vartheta(\mathbf{z}|\mathbf{x})$, and computing the gradient $\nabla_\theta \log p_\theta(\mathbf{x}|\mathbf{z})$ for the sampled $\mathbf{z}$. This gives a stochastic gradient, i.e., an unbiased estimate for the gradient with respect to the parameter $\boldsymbol{\theta}$.

Now consider computing the gradient of the ELBO with respect to the parameter ϑ, i.e.,

$$\begin{aligned}\nabla_\vartheta \mathrm{ELBO}_{\vartheta,\theta}(\mathbf{x}) &= \nabla_\vartheta \mathbb{E}_{q_\vartheta(\mathbf{z}|\mathbf{x})}\left[\log p_\theta(\mathbf{x}|\mathbf{z})\right] \\ &\quad + \nabla_\vartheta \frac{1}{2}\left(1 + \log \sigma^2_{\vartheta,i}(\mathbf{x}) - \mu^2_{\vartheta,i}(\mathbf{x}) - \sigma^2_{\vartheta,i}(\mathbf{x})\right)\end{aligned}$$

or obtaining a corresponding stochastic gradient. The second term on the right-hand side above is straightforward to compute with standard auto-differentiation packages. However, computing the gradient $\nabla_\vartheta \mathbb{E}_{q_\vartheta(\mathbf{z}|\mathbf{x})}\left[\log p_\theta(\mathbf{x}|\mathbf{z})\right]$ or a corresponding stochastic gradient is not as easy, as we cannot exchange taking the gradient and the expectation, since the expectation also depends on $\boldsymbol{\vartheta}$.

To tackle this difficulty, Rezende et al. [RMW14], see Kingma and Welling [KW19] proposed the following so-called re-parameterization trick. The idea is to introduce a random variable $\mathbf{e} \sim \mathcal{N}(0, \mathbf{I})$ and recognize that $\mathbf{z} = \boldsymbol{\mu} + \mathrm{diag}(\boldsymbol{\sigma})\mathbf{e}$ is Gaussian with mean $\boldsymbol{\mu}$ and diagonal co-variance matrix $\mathrm{diag}(\boldsymbol{\sigma})$, as desired. With this re-parameterization we have

$$\begin{aligned}\nabla_\vartheta \mathbb{E}_{\mathbf{z}\sim q_\vartheta(\mathbf{z}|\mathbf{x})}\left[\log p_\theta(\mathbf{x}|\mathbf{z})\right] &= \nabla_\vartheta \mathbb{E}_{\mathbf{e}\sim p(\mathbf{e})}\left[\log p_\theta(\mathbf{x}|\mathbf{z} = \boldsymbol{\mu} + \mathrm{diag}(\boldsymbol{\sigma})\mathbf{e})\right] \\ &= \mathbb{E}_{\mathbf{e}\sim p(\mathbf{e})}\left[\nabla_\vartheta \log p_\theta(\mathbf{x}|\mathbf{z} = \boldsymbol{\mu} + \mathrm{diag}(\boldsymbol{\sigma})\mathbf{e})\right].\end{aligned}$$

Now we can again obtain an unbiased estimate by sampling the Gaussian $\mathbf{e}$, and computing the gradient with respect to the variable $\boldsymbol{\vartheta}$.

9.5.4 Summary of training and sampling with a variational autoencoder

To summarize, we initialize the weights of the encoder and decoder as $\boldsymbol{\vartheta}_0$ and $\boldsymbol{\theta}_0$ and run stochastic gradient to minimize the loss (9.10) by iterating the following. Draw a random training example $\mathbf{x}$ from the set of training examples. Draw $\mathbf{e}_t$ from a standard normal Gaussian and compute $\mathbf{z}_t = \boldsymbol{\mu}_{\vartheta_t}(\mathbf{x}) + \mathrm{diag}(\boldsymbol{\sigma}_{\vartheta_t}(\mathbf{x}))\mathbf{e}_t$. Perform stochastic gradient method steps with step size η, i.e., compute

$$
\begin{aligned}
\boldsymbol{\theta}_{t+1} &= \boldsymbol{\theta}_t - \eta \nabla_{\boldsymbol{\theta}=\boldsymbol{\theta}_t} \frac{1}{2\sigma^2} \|\mathbf{x} - G_{\boldsymbol{\theta}}(\mathbf{z}_t)\|_2^2 \\
\vartheta_{t+1} &= \vartheta_t - \eta \nabla_{\vartheta=\vartheta_t} \frac{1}{2\sigma^2} \|\mathbf{x} - G_{\boldsymbol{\theta}}(\boldsymbol{\mu}_{\vartheta}(\mathbf{x}) + \mathrm{diag}(\boldsymbol{\sigma}_{\vartheta}(\mathbf{x}))\mathbf{e}_t\|_2^2 \\
&\quad + \eta \nabla_{\vartheta=\vartheta_t} \sum_{i=1}^{k} \frac{1}{2} \left(1 + \log \sigma_{\vartheta,i}^2(\mathbf{x}) - \mu_{\vartheta,i}^2(\mathbf{x}) - \sigma_{\vartheta,i}^2(\mathbf{x})\right).
\end{aligned}
$$

The gradients can be computed for a given encoder and decoder architecture with standard auto-differentiation software packages.

9.6 Generative adversarial networks

Generative adversarial networks [Goo+14] are a class of generative models that are trained in a game-theoretic adversarial way. They consist of a generator $G: \mathbb{R}^k \to \mathbb{R}^n$ in the form of a neural network that maps a k-dimensional latent space to the n-dimensional signal space. Once trained, $G(\mathbf{z})$ is a sample from the learned data distribution, where the latent vector $\mathbf{z}$ is from a Gaussian distribution $\mathcal{N}(0, \mathbf{I})$. For imaging applications, the generator is typically a convolutional network.

An adversarially trained model is trained with the help of a discriminator network $D: \mathbb{R}^n \to [0, 1]$ that takes as an input an image and maps it to a probability quantifying whether the input image is real or fake. The method is illustrated in Figure 9.2.

The generator and discriminator networks are trained simultaneously as follows. The goal of the discriminator is to distinguish real images from fake images, and the goal of the generator is to generate images that are indistinguishable from the real images. Say real images (i.e., images generated by the true data distribution $p(\mathbf{x})$) are labelled by $y = 1$ and fake images (i.e., those generated as $G_{\boldsymbol{\theta}}(\mathbf{z}), \mathbf{z} \sim \mathcal{N}(0, \mathbf{I})$) are labelled as $y = 0$. With this notation, the cross-entropy loss associated with a real image ($y = 1$) is

$$
-y \log D(\mathbf{x}) - (1 - y) \log(1 - D(\mathbf{x})) = -\log D(\mathbf{x}),
$$

and the cross-entropy loss associated with a false image ($y = 0$) is

$$
-y \log D(G(\mathbf{z})) - (1 - y) \log(1 - D(G(\mathbf{z}))) = -\log(1 - D(G(\mathbf{z}))).
$$

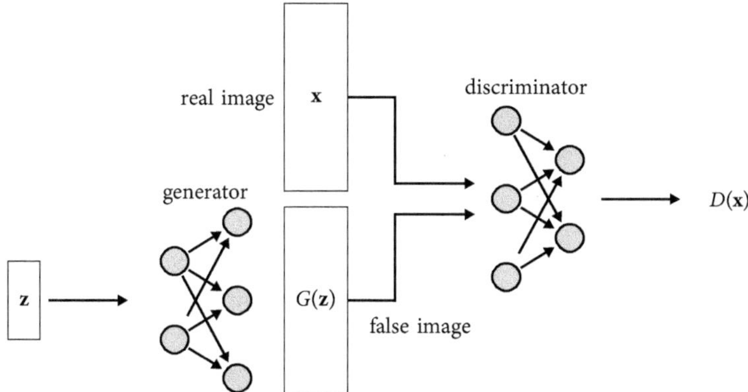

Figure 9.2 A fake image is generated by sampling a latent vector $\mathbf{z}$ from a Gaussian and evaluating the generator $G(\mathbf{z})$. The discriminator's goal is to distinguish a real image from a false image.

Suppose we train the discriminator on a dataset containing equally many true and false images. Then the loss associated with such a dataset is

$$J(\boldsymbol{\theta}, \boldsymbol{\vartheta}) = -\frac{1}{2}\mathbb{E}_{\mathbf{x}\sim p}\left[\log D_{\boldsymbol{\vartheta}}(\mathbf{x})\right] - \frac{1}{2}\mathbb{E}_{\mathbf{z}\sim\mathcal{N}(0,\mathbf{I})}\left[\log(1 - D_{\boldsymbol{\vartheta}}(G_{\boldsymbol{\theta}}(\mathbf{z})))\right], \tag{9.11}$$

where $\boldsymbol{\theta}$ and $\boldsymbol{\vartheta}$ are the parameters of the generator and the discriminator.

The discriminator wants to minimize this loss to distinguish real from fake images, and the generator wants to maximize this loss to make real and fake images indistinguishable, which leads to the max-min game

$$\max_{\boldsymbol{\theta}} \min_{\boldsymbol{\vartheta}} J(\boldsymbol{\theta}, \boldsymbol{\vartheta}).$$

The generator and discriminator are optimized via standard first-order methods, such as gradient descent: For a fixed generator, we first take a few gradient descent steps for minimizing the objective $J(\boldsymbol{\theta}, \boldsymbol{\vartheta})$, and then we update the generator by taking a gradient ascent step to maximize the objective $J(\boldsymbol{\theta}, \boldsymbol{\vartheta})$. Those updates are iterated. In practice, optimizers such as Adam are popular to carry out the optimization.

For images, a classical architecture is the DCGAN architecture [RMC16], which uses a convolutional generator (similar to the generator part of the U-net we have seen earlier), and uses a convolutional discriminator as well. There are also other ways to set up the cost function and to train a GAN, for example, Wasserstein GANs [ACB17]. GANs can generate excellent and realistic looking images, see for example the StyleGAN variants.

10
Diffusion models

Diffusion models are probabilistic models that learn a data distribution $p(\mathbf{x})$ by gradually denoising a Gaussian random variable. Diffusion models consist of a forward diffusion process that gradually adds noise to the input and a reverse denoising process that is aimed at inverting the diffusion process. See Figure 10.1 for an illustration.

Diffusion models were introduced by Sohl-Dickstein et al. [Soh+15] and became very popular when Ho et al. [HJA20] and Nichol and Dhariwal [ND21] demonstrated that diffusion models enable sampling realistic images on par with or better than other generative models, like GANs and VAEs.

10.1 Denoising diffusion probabilistic models

We start with a brief overview of diffusion models and a commonly used training and sampling procedure by Ho. et al. [HJA20] called denoising diffusion probabilistic model (DDPM). Different training and sampling procedures and parameterizations of diffusion models exist, but the DDPM parameterization is popular and is used in several subsequent works including Kingma et al. (2021) [Kin+21].

The forward diffusion process generates an output $\mathbf{x}_T$ that is nearly isotropic Gaussian (i.e., $\mathcal{N}(0, \mathbf{I})$-distributed) in T steps from a clean input image $\mathbf{x}_0$ by adding small amounts of Gaussian noise according to

$$\mathbf{x}_t = \sqrt{1 - \beta_t}\mathbf{x}_{t-1} + \sqrt{\beta_t}\epsilon_t, \quad \epsilon_t \sim \mathcal{N}(0, \mathbf{I}). \tag{10.1}$$

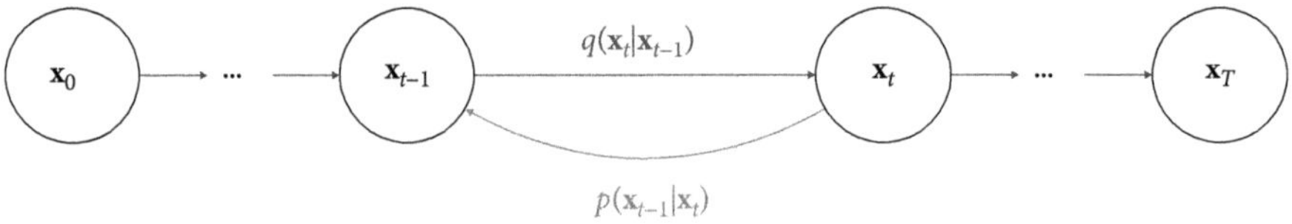

Figure 10.1 A diffusion process as a probabilistic model. The forward process (left to right) starts from a noise-free image $\mathbf{x}_0$ and gradually adds noise by sampling from the conditional distribution $q(\mathbf{x}_t|\mathbf{x}_{t-1})$, which adds small amounts of Gaussian noise. The backward process (right to left) aims at reverting the forward process by sampling from the conditional distribution $p(\mathbf{x}_{t-1}|\mathbf{x}_t)$, which aims at denoising.

Deep Learning for Computational Imaging. Reinhard Heckel, Oxford University Press. © Reinhard Heckel (2025).
DOI: 10.1093/oso/9780198947172.003.0010

Here, $\beta_1 < \beta_2 < \cdots < \beta_T$ are step sizes chosen so the updates become larger when the sample becomes noisier.

Given the clean image $\mathbf{x}_0$, we can sample a point at an arbitrary time step in the diffusion process, $\mathbf{x}_t$, according to the conditional distribution

$$q(\mathbf{x}_t|\mathbf{x}_0) = \mathcal{N}\left(\sqrt{1-\sigma_t^2}\mathbf{x}_0, \sigma_t^2\mathbf{I}\right), \tag{10.2}$$

where we defined the variance $\sigma_t^2 = 1 - \prod_{s=1}^{t}(1-\beta_s)$. See the end of the section on how equation (10.2) follows from equation (10.1).

Denote the signal-to-noise ratio of the point $\mathbf{x}_t$ as $\text{SNR}(t) = \frac{1-\sigma_t^2}{\sigma_t^2}$. The step-size schedule β_t, or equivalently the noise-schedule σ_t^2, is designed so that $\sigma_t^2 \to 1$, and thus the distribution of the output conditioned on the input, $q(\mathbf{x}_T|\mathbf{x}_0)$, is (approximately) Gaussian with zero-mean and unit co-variance matrix.

See Figure 10.2 for a few samples from a diffusion process applied to an image.

In order to generate a sample $\mathbf{x}_0$ with a diffusion process, we wish to reverse the diffusion process by starting with sampling a standard Gaussian random vector $\mathbf{x}_T \sim \mathcal{N}(0, \mathbf{I})$ and generating the states $\mathbf{x}_{T-1}, \mathbf{x}_{T-2}, \ldots, \mathbf{x}_0$ consecutively by sampling according to the conditional density $p(\mathbf{x}_{t-1}|\mathbf{x}_t)$. The conditional density $p(\mathbf{x}_{t-1}|\mathbf{x}_t)$ is modelled as a Gaussian, and parameterized with a neural network ϵ_θ with parameters $\boldsymbol{\theta}$ that are learned based on data.

Sampling from the conditional distribution is accomplished by computing the prediction

$$\mathbf{x}_{t-1} = \frac{1}{\sqrt{1-\beta_t}}\left(\mathbf{x}_t - \frac{\beta_t}{\sigma_t}\epsilon_\theta(\mathbf{x}_t, t)\right) + \tilde{\sigma}_t\mathbf{z}, \quad \mathbf{z} \sim \mathcal{N}(0, \mathbf{I}), \tag{10.3}$$

where ϵ_θ is a neural network that predicts the noise ϵ which when added to the image yields the state $\mathbf{x}_t$, and where $\tilde{\sigma}_t^2 = \frac{\sigma_{t-1}^2}{\sigma_t^2}\beta_t$. The neural network is time-conditioned, and it is common to use a time-conditional U-net, see Figure 10.3 for an illustration.

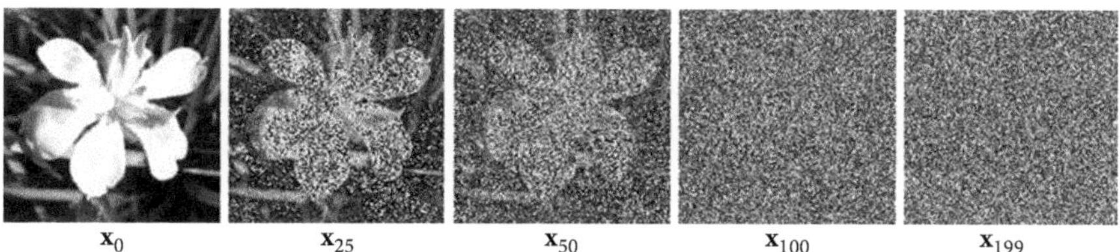

Figure 10.2 Illustration of a diffusion process with $T = 200$ steps and a cosine-schedule for the parameter β_t. The samples are obtained by gradually adding noise or directly via equation (10.2). The noise variances are $\sigma_0^2 = 0, \sigma_{25}^2 = 0.045, \sigma_{50}^2 = 0.1586, \sigma_{100}^2 = 0.5139, \sigma_{199}^2 = 1$.

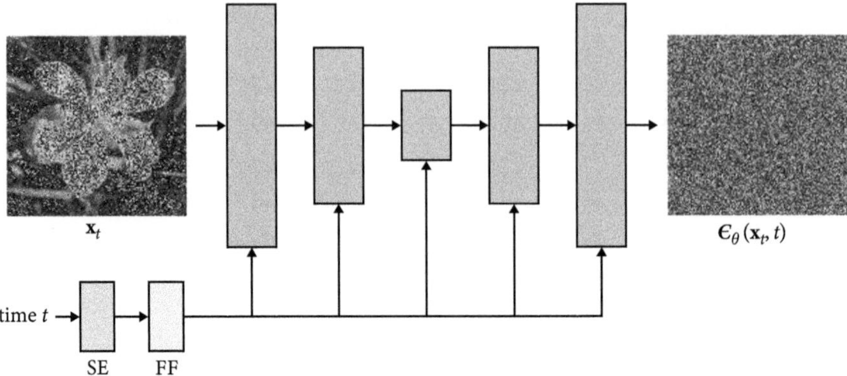

Figure 10.3 The time-conditional network $\epsilon_\theta(\mathbf{x}_t, t)$ predicts the noise which when added to the image yields $\mathbf{x}_t$. The architecture used for the time-conditional network is typically a U-net. The time-conditioning is implemented as follows: First a sinusoidal embedding is used that transforms the time to $\mathrm{SE}(t) = [\sin(a_1 t), \sin(a_2 t), \ldots, \sin(a_k t)]$, where $a_1, \ldots, a_k$ are different frequency parameters. This embedding of the time is then transformed using a simple feed-forward network (FF) to an embedding that is integrated into each layer of the U-net. The integration into the U-net happens by conditioning normalization layers on those parameters, or through addition to the feature maps, or through other mechanisms.

The neural network ϵ_θ is trained by minimizing the objective

$$L(\boldsymbol{\theta}) = \mathbb{E}_{\mathbf{x}_0, t, \epsilon}\left[\left\|\epsilon - \epsilon_\theta(\sqrt{1 - \sigma_t^2}\mathbf{x}_0 + \sigma_t \epsilon, t)\right\|_2^2\right], \tag{10.4}$$

where $\mathbf{x}_0$ is drawn from the image distribution q, the time step t is drawn i.i.d. uniformly from $\{1, \ldots, T\}$, and the noise $\epsilon \sim \mathcal{N}(0, \mathbf{I})$ is standard Gaussian.

In practice, the objective (10.4) is minimized by running the stochastic gradient method. Specifically, we sample an image $\mathbf{x}_0$ from the set of training images, as well as a time step t and noise vector ϵ, and then update the network ϵ_θ by performing a gradient update in the direction of the stochastic gradient

$$\nabla_\theta \left\|\epsilon - \epsilon_\theta\left(\sqrt{1 - \sigma_t^2}\mathbf{x}_0 + \sigma_t \epsilon, t\right)\right\|_2^2.$$

10.1.1 Detailed discussion of the loss and DDPM parameterization

We next discuss the parameterization and motivation for the loss function, as well as the sampling approach in more detail.

Recall that in order to generate a sample $\mathbf{x}_0$ based on a diffusion process, we wish to reverse the diffusion process by starting with sampling a standard Gaussian random vector $\mathbf{x}_T$ and generating the iterates $\mathbf{x}_{T-1}, \mathbf{x}_{T-2}, \dots, \mathbf{x}_0$ consecutively by sampling from the conditional distribution $p(\mathbf{x}_{t-1}|\mathbf{x}_t)$. The conditional distribution $p(\mathbf{x}_{t-1}|\mathbf{x}_t)$ is modelled as Gaussians with parameters $\boldsymbol{\theta}$ learned from data. The DDPM parameterization used by Ho et al. [HJA20] (and by subsequent works [Kin+21]) is

$$p_{\theta}(\mathbf{x}_{t-1}|\mathbf{x}_t) = \mathcal{N}(\boldsymbol{\mu}_{\theta}(\mathbf{x}_t, t), \boldsymbol{\Sigma}_{\theta}(\mathbf{x}_t, t)),$$

where the mean $\boldsymbol{\mu}_{\theta}(\mathbf{x}_t, t)$ is trainable, and the co-variance matrix (or its diagonal) $\boldsymbol{\Sigma}_{\theta}(\mathbf{x}_t, t)$ can be trainable as well, but Ho et al. [HJA20] achieved the best results by fixing it to $\beta_t\mathbf{I}$.

To train, we can optimize the evidence lower bound (ELBO) on the negative log-likelihood (see equation (9.4) for the ELBO):

$$\mathbb{E}[-\log p_{\theta}(\mathbf{x})] \le \mathbb{E}_q\left[-\log \frac{p_{\theta}(\mathbf{x}_0, \dots, \mathbf{x}_T)}{q(\mathbf{x}_1, \dots, \mathbf{x}_T|\mathbf{x}_0)}\right].$$

Re-writing the right-hand side of the ELBO further yields

$$\begin{aligned}
&\mathbb{E}_q\left[-\log \frac{p_{\theta}(\mathbf{x}_0, \dots, \mathbf{x}_T)}{q(\mathbf{x}_1, \dots, \mathbf{x}_T|\mathbf{x}_0)}\right] \\
&= \mathbb{E}_q\left[-\log \frac{p_{\theta}(\mathbf{x}_T)\prod_{t=1}^{T} p_{\theta}(\mathbf{x}_{t-1}|\mathbf{x}_t)}{\prod_{t=1}^{T} q(\mathbf{x}_t|\mathbf{x}_{t-1})}\right] \\
&= \mathbb{E}_q\left[-\log p_{\theta}(\mathbf{x}_T) - \sum_{t=2}^{T} \log \frac{p_{\theta}(\mathbf{x}_{t-1}|\mathbf{x}_t)}{q(\mathbf{x}_t|\mathbf{x}_{t-1})} - \log \frac{p_{\theta}(\mathbf{x}_0|\mathbf{x}_1)}{q(\mathbf{x}_1|\mathbf{x}_0)}\right] \\
&\overset{\text{(i)}}{=} \mathbb{E}_q\left[-\log p_{\theta}(\mathbf{x}_T) - \sum_{t=2}^{T} \log \frac{p_{\theta}(\mathbf{x}_{t-1}|\mathbf{x}_t)q(\mathbf{x}_{t-1}|\mathbf{x}_0)}{q(\mathbf{x}_{t-1}|\mathbf{x}_t, \mathbf{x}_0)q(\mathbf{x}_t|\mathbf{x}_0)} - \log \frac{p_{\theta}(\mathbf{x}_0|\mathbf{x}_1)}{q(\mathbf{x}_1|\mathbf{x}_0)}\right] \\
&= \mathbb{E}_q\Bigg[-\log p_{\theta}(\mathbf{x}_T) - \sum_{t=2}^{T} \log \frac{p_{\theta}(\mathbf{x}_{t-1}|\mathbf{x}_t)}{q(\mathbf{x}_{t-1}|\mathbf{x}_t, \mathbf{x}_0)} + \log q(\mathbf{x}_{t-1}|\mathbf{x}_0) - \log q(\mathbf{x}_t|\mathbf{x}_0) \\
&\quad - \log p_{\theta}(\mathbf{x}_0|\mathbf{x}_1) - \log q(\mathbf{x}_1|\mathbf{x}_0)\Bigg] \\
&\overset{\text{(ii)}}{=} \mathbb{E}_q\left[-\log \frac{p_{\theta}(\mathbf{x}_T)}{q(\mathbf{x}_T|\mathbf{x}_0)} - \sum_{t=2}^{T} \log \frac{p_{\theta}(\mathbf{x}_{t-1}|\mathbf{x}_t)}{q(\mathbf{x}_{t-1}|\mathbf{x}_t, \mathbf{x}_0)} - \log p_{\theta}(\mathbf{x}_0|\mathbf{x}_1)\right] \\
&\overset{\text{(iii)}}{=} \mathbb{E}_q\left[-\log \frac{p_{\theta}(\mathbf{x}_T)}{q(\mathbf{x}_T|\mathbf{x}_0)} - \sum_{t=1}^{T} \log \frac{p_{\theta}(\mathbf{x}_{t-1}|\mathbf{x}_t)}{q(\mathbf{x}_{t-1}|\mathbf{x}_t, \mathbf{x}_0)}\right],
\end{aligned}$$

where equation (i) follows by using that $\mathbf{x}_t$ and $\mathbf{x}_0$ are conditionally independent, given $\mathbf{x}_{t-1}$ and from an application of Bayes' rule, i.e.,

$q(\mathbf{x}_t|\mathbf{x}_{t-1}) = q(\mathbf{x}_t|\mathbf{x}_{t-1}, \mathbf{x}_0) = q(\mathbf{x}_{t-1}|\mathbf{x}_t, \mathbf{x}_0)q(\mathbf{x}_t|\mathbf{x}_0)/q(\mathbf{x}_{t-1}|\mathbf{x}_0)$. Moreover, equation (ii) follows from re-arranging the terms, and equation (iii) follows from $q(\mathbf{x}_0|\mathbf{x}_1, \mathbf{x}_0) = 1$.

Thus, we have the bound

$$\mathbb{E}\left[-\log p_\theta(\mathbf{x}_0)\right] \leq D_{KL}(q(\mathbf{x}_T|\mathbf{x}_0) \parallel p_\theta(\mathbf{x}_T)) + \sum_{t=1}^{T} L_{t-1}, \tag{10.5}$$

where we defined

$$L_{t-1} = \mathbb{E}_{q(\mathbf{x}_t|\mathbf{x}_0)}\left[D_{KL}(q(\mathbf{x}_{t-1}|\mathbf{x}_t, \mathbf{x}_0) \parallel p_\theta(\mathbf{x}_{t-1}|\mathbf{x}_t))\right],$$

and were we used that $-\log p_\theta(\mathbf{x}_0|\mathbf{x}_1)$ is non-negative. Moreover, D_{KL} is the Kullback-Leibler (KL) divergence defined earlier in equation (9.5).

The prior $p_\theta(\mathbf{x}_T)$ is usually chosen as a standard normal distribution, and thus the first term in the bound (10.5) is constant and thus not optimized over.

Regarding the other terms; with Bayes' rule the true posterior conditioned on $\mathbf{x}_0$ can be written as

$$q(\mathbf{x}_{t-1}|\mathbf{x}_t, \mathbf{x}_0) = \mathcal{N}(\tilde{\boldsymbol{\mu}}(\mathbf{x}_t, \mathbf{x}_0), \tilde{\sigma}_t^2\mathbf{I}), \tag{10.6}$$

where

$$\tilde{\boldsymbol{\mu}}(\mathbf{x}_t, \mathbf{x}_0) = \sqrt{1 - \beta_t \frac{\sigma_{t-1}^2}{\sigma_t^2}}\mathbf{x}_t + \beta_t \frac{\sqrt{1 - \sigma_{t-1}^2}}{\sigma_t^2}\mathbf{x}_0, \quad \tilde{\sigma}_t^2 = \frac{\sigma_{t-1}^2}{\sigma_t^2}\beta_t. \tag{10.7}$$

The distribution $p_\theta(\mathbf{x}_{t-1}|\mathbf{x}_t)$ is usually parameterized with the true posterior $q(\mathbf{x}_{t-1}|\mathbf{x}_t, \mathbf{x}_0)$ with the unknown $\mathbf{x}_0$ replaced by an estimate $\mathbf{x}_\theta(\mathbf{x}_t, t)$ based on the current iterate $\mathbf{x}_t$. This gives

$$p_\theta(\mathbf{x}_{t-1}|\mathbf{x}_t) = \mathcal{N}(\tilde{\boldsymbol{\mu}}(\mathbf{x}_t, \mathbf{x}_\theta(\mathbf{x}_t, t)), \tilde{\sigma}_t^2\mathbf{I}). \tag{10.7}$$

With this, the term L_{t-1} becomes

$$\begin{aligned} L_{t-1} &= \mathbb{E}_{q(\mathbf{x}_t|\mathbf{x}_0)}\left[D_{KL}(q(\mathbf{x}_{t-1}|\mathbf{x}_t, \mathbf{x}_0) \parallel p_\theta(\mathbf{x}_{t-1}|\mathbf{x}_t))\right] \\ &\propto \mathbb{E}_{q(\mathbf{x}_t|\mathbf{x}_0)}\left[\frac{1}{\tilde{\sigma}_t^2}\|\tilde{\boldsymbol{\mu}}(\mathbf{x}_{t-1}, \mathbf{x}_0) - \tilde{\boldsymbol{\mu}}(\mathbf{x}_{t-1}, \mathbf{x}_\theta(\mathbf{x}_t, t))\|_2^2\right] \\ &= \mathbb{E}_{q(\mathbf{x}_t|\mathbf{x}_0)}\left[\frac{\beta_t(1 - \sigma_{t-1}^2)}{\sigma_t^2\sigma_{t-1}^2}(\mathrm{SNR}(t-1) - \mathrm{SNR}(t))\,\|\mathbf{x}_0 - \mathbf{x}_\theta(\mathbf{x}_t, t)\|_2^2\right], \end{aligned}$$

where the last equality follows from Equation (10.7). In the second equation the proportional to sign ($\propto$) means equality up to a numerical constant.

Recall that $\mathbf{x}_\theta(\mathbf{x}_t, t)$ predicts the image $\mathbf{x}_0$ based on $\mathbf{x}_t$. From Equation (10.2), we have that $\mathbf{x}_t = \sqrt{1-\sigma_t^2}\mathbf{x}_0 + \sigma_t \epsilon$. Thus, instead of predicting the image $\mathbf{x}_0$ we can predict the noise ϵ which when added to the image yields the observation $\mathbf{x}_t$. Predicting the noise and re-parameterizing according to

$$\epsilon_\theta(\mathbf{x}_t, t) = \frac{\mathbf{x}_t - \sqrt{1-\sigma_t^2}\mathbf{x}_\theta(\mathbf{x}_t, t)}{\sigma_t} \tag{10.9}$$

yields the loss

$$\mathbb{E}_{\mathbf{x}_0,\epsilon,t}\left[(\mathrm{SNR}(t-1) - \mathrm{SNR}(t))\,\mathrm{SNR}(t)\left\|\epsilon - \epsilon_\theta(\sqrt{1-\sigma_t^2}\mathbf{x}_0 + \sigma_t\epsilon, t)\right\|_2^2\right]. \tag{10.10}$$

Predicting the noise instead of the signal $\mathbf{x}_0$ is considered to yield better performance.

Instead of the loss (10.10) Ho et al. [HJA20] and subsequent works used the loss function

$$\mathbb{E}_{\mathbf{x}_0,t,\epsilon}\left[\left\|\epsilon - \epsilon_\theta(\sqrt{1-\sigma_t^2}\mathbf{x}_0 + \sigma_t\epsilon, t)\right\|_2^2\right],$$

stated previously is Equation (10.4) in the overview of diffusion models.

Sampling.

In order to sample, we start from a Gaussian random vector $\mathbf{x}_T$ and iteratively sample $\mathbf{x}_{t-1}$ from the Gaussian distribution (10.6) conditioned on $\mathbf{x}_t$ and conditioned on an estimate of the clean image $\mathbf{x}_0$. As an estimate for the clean image $\mathbf{x}_0$, we use the estimate

$$\mathbf{x}_\theta(\mathbf{x}_t, t) = \frac{1}{\sqrt{1-\sigma_t^2}}(\mathbf{x}_t - \sigma_t\epsilon_\theta(\mathbf{x}_t, t)),$$

which is based on the noise-predicting neural network $\epsilon_\theta(\mathbf{x}_t, t)$ we trained (see Equation (10.9)). Using this estimate for the clean image in the formula for the mean (10.7) yields the mean

$$\begin{aligned}\tilde{\mu}(\mathbf{x}_t, \mathbf{x}_\theta(\mathbf{x}_t, t)) &= \sqrt{1-\beta_t}\frac{\sigma_{t-1}^2}{\sigma_t^2}\mathbf{x}_t + \beta_t\frac{\sqrt{1-\sigma_{t-1}^2}}{\sigma_t^2\sqrt{1-\sigma_t^2}}(\mathbf{x}_t - \sigma_t\epsilon_\theta(\mathbf{x}_t, t)) \\ &= \frac{1}{\sqrt{1-\beta_t}}\left(\mathbf{x}_t - \frac{\beta_t}{\sigma_t}\epsilon_\theta(\mathbf{x}_t, t)\right),\end{aligned}$$

where the last equality follows from simplification using that by definition of σ_t, $\frac{\sqrt{1-\sigma_{t-1}^2}}{\sqrt{1-\sigma_t^2}} = \frac{1}{\sqrt{1-\beta_t}}$ as well as using that $\sigma_t^2 = (1-\beta_t)\sigma_{t-1}^2 + \beta_t$.

Thus the prediction in Equation (10.3), i.e.,

$$\mathbf{x}_{t-1} = \frac{1}{\sqrt{1-\beta_t}}\left(\mathbf{x}_t - \frac{\beta_t}{\sigma_t}\epsilon_\theta(\mathbf{x}_t, t)\right) + \tilde{\sigma}_t \mathbf{z}, \quad \mathbf{z} \sim \mathcal{N}(0, \mathbf{I}),$$

amounts to sampling from the distribution $q(\mathbf{x}_{t-1}|\mathbf{x}_t, \mathbf{x}_\theta(\mathbf{x}_t, t)) = \mathcal{N}(\tilde{\boldsymbol{\mu}}(\mathbf{x}_t, \mathbf{x}_\theta(\mathbf{x}_t, t)), \tilde{\sigma}_t^2 \mathbf{I})$, as desired.

Proof of Equation (10.2): *Note that*

$$\begin{aligned}
\mathbf{x}_t &= \sqrt{1-\beta_t}\mathbf{x}_{t-1} + \sqrt{\beta_t}\mathbf{z}_{t-1} \\
&= \sqrt{1-\beta_t}(\sqrt{1-\beta_{t-1}}\mathbf{x}_{t-2} + \sqrt{\beta_{t-1}}\mathbf{z}_{t-2}) + \sqrt{\beta_t}\mathbf{z}_{t-1} \\
&= \sqrt{1-\beta_t}\sqrt{1-\beta_{t-1}}\mathbf{x}_{t-1} + \sqrt{1-\beta_t}\sqrt{\beta_{t-1}}\mathbf{z}_{t-2} + \sqrt{\beta_t}\mathbf{z}_{t-1} \\
&\overset{(i)}{=} \sqrt{(1-\beta_t)(1-\beta_{t-1})}\mathbf{x}_{t-2} + \sqrt{1-(1-\beta_t)(1-\beta_{t-1})}\mathbf{z}_{t-2} \\
&= \sqrt{\prod_{s=1}^{t}(1-\beta_s)}\mathbf{x}_0 + \sqrt{1-\prod_{s=1}^{t}(1-\beta_s)}\mathbf{z}_0.
\end{aligned}$$

Here, equation (i) follows from the sum of two Gaussian $z_1 \sim \mathcal{N}(0, \sigma_1^2), z_2 \sim \mathcal{N}(0, \sigma_2^2)$ *being Gaussian with variance* $z_1 + z_2 \sim \mathcal{N}(0, \sigma_1^2 + \sigma_2^2)$.

Proof of Equation (10.6): *By Bayes' theorem, we have*

$$q(\mathbf{x}_{t-1}|\mathbf{x}_t, \mathbf{x}_0) = q(\mathbf{x}_t|\mathbf{x}_{t-1}, \mathbf{x}_0)\frac{q(\mathbf{x}_{t-1}, \mathbf{x}_0)}{q(\mathbf{x}_t, \mathbf{x}_0)}. \tag{10.11}$$

Note that the term $q(\mathbf{x}_t, \mathbf{x}_0)$ *does not depend on* $\mathbf{x}_{t-1}$. *Using in addition Equations (10.1) and (10.2), the density is proportional to*

$$q(\mathbf{x}_{t-1}|\mathbf{x}_t, \mathbf{x}_0) \propto \exp\left(-\frac{\|\mathbf{x}_t - \sqrt{1-\beta_t}\mathbf{x}_{t-1}\|_2^2}{2\beta_t} - \frac{\left\|\mathbf{x}_{t-1} - \sqrt{1-\sigma_{t-1}^2}\mathbf{x}_0\right\|_2^2}{2\sigma_{t-1}^2}\right)$$

$$= \exp\left(\frac{-\|\mathbf{x}_{t-1}\|_2^2((1-\beta_t)\sigma_{t-1}^2 + \beta_t) - 2\mathbf{x}_{t-1}^T\left(\sqrt{1-\beta_t}\sigma_{t-1}^2\mathbf{x}_t + \sqrt{1-\sigma_{t-1}^2}\beta_t\mathbf{x}_t\right)}{2\sigma_{t-1}^2\beta_t}\right)$$

$$= \exp\left(-\frac{-\|\mathbf{x}_{t-1}\|_2^2 - 2\mathbf{x}_{t-1}^T\left(\sqrt{1-\beta_t}\frac{\sigma_{t-1}^2}{\sigma_t^2}\mathbf{x}_t + \frac{\sqrt{1-\sigma_{t-1}^2}}{\sigma_t^2}\beta_t\mathbf{x}_t\right)}{2\frac{\sigma_{t-1}^2}{\sigma_t^2}\beta_t}\right)$$

where the last equation follows from $\sigma_t^2 = (1-\beta_t)\sigma_{t-1}^2 + \beta_t$. From the last equation, it follows that the density $q(\mathbf{x}_{t-1}|\mathbf{x}_t, \mathbf{x}_0)$ is Gaussian with mean and variance given by Equation (10.7). This concludes the proof of Equation (10.6).

10.2 Score-based generative models

Diffusion models are closely related to score-based generative models [SE19]. Score-based models learn a model for an unknown data distribution $p(\mathbf{x})$ based on examples $\{\mathbf{x}_1, ..., \mathbf{x}_N\}$ drawn i.i.d. from the distribution $p(\mathbf{x})$, so that we can generate new data that looks like data from the distribution $p(\mathbf{x})$, like other classes of generative models.

Score-based models are based on parameterizing the score function with a neural network. The score function of a probability distribution with probability density p is a vector-valued function defined as the gradient of the log-probability density

$$\mathbf{s}(\mathbf{x}) = \nabla_{\mathbf{x}} \log p(\mathbf{x}). \tag{10.12}$$

From the score function, the density can be computed. Specifically, integration yields the log-likelihood of the density up to a constant, i.e., $\log p(\mathbf{x}) + C$, and taking the exponential yields the density times a constant, i.e., $p(\mathbf{x}) \cdot \exp(C)$. From $p(\mathbf{x}) \cdot \exp(C)$, the probability density p can be computed by using that the density integrates to one.

As an example, for a multi-variate Gaussian distribution with mean $\boldsymbol{\mu} \in \mathbb{R}^n$ and unit co-variance matrix, i.e., $p(\mathbf{x}) = \frac{1}{(2\pi)^{n/2}} \exp\left(-\frac{1}{2}\|\mathbf{x} - \boldsymbol{\mu}\|_2^2\right)$, the score function is $\mathbf{s}(\mathbf{x}) = \boldsymbol{\mu} - \mathbf{x}$, which is a vector pointing from $\mathbf{x}$ towards the mean $\boldsymbol{\mu}$. Figure 10.4 illustrates the score function for the mixture of two Gaussians.

Computationally the score function can be easier to work with, since unlike the density, the score function does not need to be normalized, and therefore can be parameterized directly by a neural network. If we parameterize the density $p(\mathbf{x})$ with a neural network, we have to ensure that it is a density function, i.e., that it integrates to one or approximately to one, for example with a normalizing constant. Modelling the score function instead of the density function sidesteps this issue since the score is independent of a normalizing constant.

In order to train a score-based model, in principle, a score function $s_\theta(\mathbf{x})$ parameterized by a neural network with parameters $\boldsymbol{\theta}$ can be fitted to a data distribution so that score function is approximately equal to the data-density $p(\mathbf{x})$ (i.e., $s_\theta(\mathbf{x}) \approx \nabla_{\mathbf{x}} \log p(\mathbf{x})$). This can be done by minimizing the Fisher divergence between the model and the data distribution, defined as

$$\frac{1}{2}\mathbb{E}_{\mathbf{x}\sim p}\left[\|\nabla_{\mathbf{x}} \log p(\mathbf{x}) - s_\theta(\mathbf{x})\|_2^2\right].$$

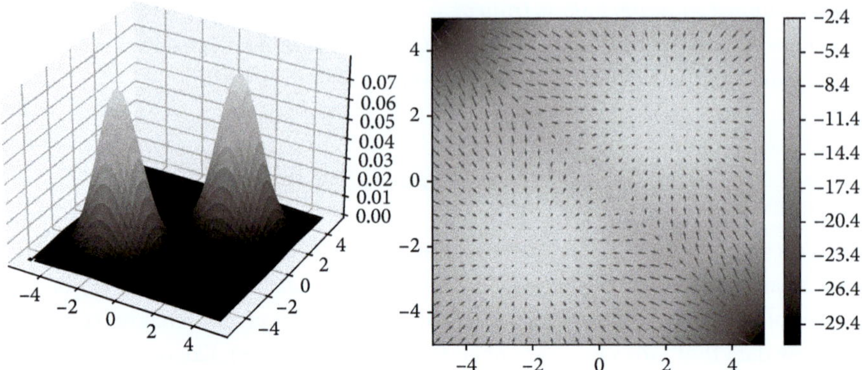

Figure 10.4 The probability density **(left)** of a mixture of two Gaussians, along with score function (gradient vectors) on top of the log-probability density **(right)**.

However, the Fisher divergence cannot be directly computed since the score function $\nabla_{\mathbf{x}} \log p(\mathbf{x})$ is unknown, and we are only given examples $\{\mathbf{x}_1, \ldots, \mathbf{x}_N\}$ from the training distribution. Moreover, as observed by Song and Ermon [SE19], such standard score matching fails to learn good estimates of the score function in low-density regions of the distribution. Both issues can be addressed by learning a score function with denoising score matching, as discussed below.

10.2.1 Sampling with Langevin dynamics

Suppose we are given a score function of a probability distribution with density p, and our goal is to sample from this probability distribution. Langevin dynamics enables sampling from the probability distribution based on the score function $\mathbf{s}(\mathbf{x}) = \nabla_{\mathbf{x}} \log p(\mathbf{x})$. Langevin dynamics starts from an initial value $\mathbf{x}_0$ drawn randomly from a prior distribution (often Gaussian) and iterates

$$\mathbf{x}_{t+1} = \mathbf{x}_t + \frac{\alpha}{2} \nabla_{\mathbf{x}} \log p(\mathbf{x}_t) + \sqrt{\alpha} \epsilon_t, \quad \epsilon_t \sim \mathcal{N}(0, \mathbf{I}), \tag{10.13}$$

where α is a step-size parameter. For the step-size parameter going to zero ($\alpha \to 0$) and the number of steps going to infinity ($T \to \infty$), we are guaranteed to sample from the probability distribution with density $p(\mathbf{x})$.

In practice, we are given an estimate of the score function parameterized by an neural network, $\mathbf{s}_\theta(\mathbf{x})$, and use the estimate of the score function instead of the score function to run Langevin dynamics.

10.2.2 Denoising score-matching

For image generation, Song and Ermon [SE19] observed that simply using a score function estimated with standard score-matching techniques and sampling with Langevin dynamics does not work well, since the estimated score functions tend to be inaccurate in low-density regions of the probability distribution. That is because in low-density regions, little data is available for computing the score-matching objective, and thus the score tends to be inaccurate in those regions.

To alleviate this issue, Song and Ermon [SE19] proposed to train with so-called denoised score matching, a technique proposed by Vincent [Vin11]. We first explain denoised score matching, and then how it is used to train score-based models.

Define a noise-perturbed distribution

$$p_\sigma(\mathbf{x}) = \int p(\mathbf{z}) g_\sigma(\mathbf{z} - \mathbf{x}) d\mathbf{z}, \tag{10.14}$$

where g is the density of a multi-variate Gaussian with zero-mean and co-variance matrix $\sigma^2\mathbf{I}$. We can sample from the noise-perturbed distribution $p_\sigma(\mathbf{x})$ by sampling a data point from the original distribution and adding Gaussian noise $\epsilon \sim \mathcal{N}(0, \sigma^2\mathbf{I})$ to it, which yields the noisy data point $\tilde{\mathbf{x}} = \mathbf{x} + \epsilon$.

The score-matching objective for the perturbed data is

$$\mathbb{E}_{p_\sigma}\left[\|\mathbf{s}_\theta(\tilde{\mathbf{x}}) - \nabla_{\tilde{\mathbf{x}}} \log p_\sigma(\tilde{\mathbf{x}})\|_2^2\right]. \tag{10.15}$$

Vincent [Vin11] has shown that the score-matching objective for the perturbed data is

$$\mathbb{E}_{\tilde{\mathbf{x}}}\left[\|\mathbf{s}_\theta(\tilde{\mathbf{x}}) - \nabla_{\tilde{\mathbf{x}}} \log p_\sigma(\tilde{\mathbf{x}})\|_2^2\right] = \mathbb{E}_{\mathbf{x},\tilde{\mathbf{x}}}\left[\left\|\mathbf{s}_\theta(\tilde{\mathbf{x}}) + \frac{\tilde{\mathbf{x}} - \mathbf{x}}{\sigma^2}\right\|_2^2\right] + C, \tag{10.16}$$

where C is a constant independent of the parameter $\boldsymbol{\theta}$ we optimize over. Let ϵ be the noise which when added to the original image $\mathbf{x}$ produces the noisy image $\tilde{\mathbf{x}} = \mathbf{x} + \epsilon$. With this notation, the score-matching objective for the perturbed data becomes

$$\mathbb{E}_{\tilde{\mathbf{x}}}\left[\|\mathbf{s}_\theta(\tilde{\mathbf{x}}) - \nabla_{\tilde{\mathbf{x}}} \log p_\sigma(\tilde{\mathbf{x}})\|_2^2\right] = \mathbb{E}_{\mathbf{x},\epsilon \sim \mathcal{N}(0,\sigma^2\mathbf{I})}\left[\left\|\mathbf{s}_\theta(\mathbf{x} + \epsilon) + \frac{\epsilon}{\sigma^2}\right\|_2^2\right] + C.$$

Thus, the score predicts the (scaled) noise when added to the original image $\mathbf{x}$ yields the noisy image $\tilde{\mathbf{x}}$.

Denoised score matching perturbs the data points by adding noise to the data points during training. The larger the noise, the more the low-density regions are covered, but the further the learned score will be from the original distribution.

To balance those two effects, Song and Ermon [SE19; SE20] proposed to train on multiple scale of noise perturbation simultaneously. Specifically, a noise-conditioned score function $\mathbf{s}_\theta(\mathbf{x}, \sigma_t)$ that approximates the noise-perturbed distribution, i.e., $\mathbf{s}_\theta(\mathbf{x}, \sigma_i) \approx \nabla_\mathbf{x} \log p_{\sigma_i}(\mathbf{x}), i = 1, \ldots, L$ is trained with denoised score matching. The model is then trained by minimizing a weighted sum of Fisher divergences

$$\sum_i \sigma_i^2 \mathbb{E}_{\tilde{\mathbf{x}} \sim p_{\sigma_i}} \left[\|\mathbf{s}_\theta(\tilde{\mathbf{x}}, \sigma_i) - \nabla_{\tilde{\mathbf{x}}} \log p_{\sigma_i}(\tilde{\mathbf{x}})\|_2^2 \right]. \tag{10.17}$$

The weighting factor σ_i^2 was chosen by Song and Ermon [SE19] for the following reason. If the score function is trained to optimality, then the norm $\|\mathbf{s}_\theta(\mathbf{x}, \sigma)\|_2$ is proportional to $1/\sigma$, and thus choosing the weighting factors as σ_i^2 ensures that all losses are weighted similarly. The objective is optimized with denoised score matching, i.e., we minimize the objective

$$\sum_{i=1}^{L} \sigma_i^2 \mathbb{E}_{\mathbf{x}, \boldsymbol{\epsilon} \sim \mathcal{N}(0, \sigma_i^2 \mathbf{I})} \left[\left\| \mathbf{s}_\theta(\mathbf{x} + \boldsymbol{\epsilon}, \sigma_i) + \frac{\boldsymbol{\epsilon}}{\sigma_i^2} \right\|_2^2 \right]. \tag{10.18}$$

This objective can be optimized with the stochastic gradient method by sampling a noise scale t, an image $\mathbf{x}$ from the set of given images, and a noise vector $\boldsymbol{\epsilon}$.

The variances $\sigma_i, i = 1, \ldots, L$ and the parameterization of the noise-conditioned score network are important parameters to make score networks work, and Song and Ermon [SE19; SE20] found the following choices to be useful for generating images. Choose the variances as a geometric progression and parameterize the score-based model as $\mathbf{s}_\theta(\mathbf{x}, \sigma) = \mathbf{s}_\theta(\mathbf{x})/\sigma$, where $\mathbf{s}_\theta(\mathbf{x})$ is an unconditional score network often chosen as the U-net.

Proof of Equation (10.16): *To prove Equation (10.16), it is key to establish*

$$\mathbb{E}_{\tilde{\mathbf{x}}} \left[\|\mathbf{s}_\theta(\tilde{\mathbf{x}}) - \nabla_{\tilde{\mathbf{x}}} \log p_\sigma(\tilde{\mathbf{x}})|_2^2 \right] = \mathbb{E}_{\mathbf{x}, \tilde{\mathbf{x}}} \left[\|\mathbf{s}_\theta(\tilde{\mathbf{x}}) - \nabla_{\tilde{\mathbf{x}}} \log p(\tilde{\mathbf{x}}|\mathbf{x})\|_2^2 \right] + C. \tag{10.19}$$

From Equation (10.19), Equation (10.16) follows directly by using that the conditional distribution $p(\tilde{\mathbf{x}}|\mathbf{x})$ is multi-variate Gaussian, and thus the logarithm of the score function of the distribution can be computed as

$$\begin{aligned} \nabla_{\tilde{\mathbf{x}}} \log p(\tilde{\mathbf{x}}|\mathbf{x}) &= \nabla_{\tilde{\mathbf{x}}} \log C \exp\left(-\|\tilde{\mathbf{x}} - \mathbf{x}\|_2^2/(2\sigma^2)\right) \\ &= -\frac{\tilde{\mathbf{x}} - \mathbf{x}}{\sigma^2}. \end{aligned}$$

To establish Equation (10.19), note that

$$\mathbb{E}_{\tilde{\mathbf{x}}} \left[\|\nabla_{\tilde{\mathbf{x}}} \log p_\sigma(\tilde{\mathbf{x}}) - \mathbf{s}_\theta(\tilde{\mathbf{x}})\|_2^2 \right] = \mathbb{E}_{\mathbf{x}, \tilde{\mathbf{x}}} \left[\|\mathbf{s}_\theta(\tilde{\mathbf{x}})\|_2^2 \right] + \mathbb{E}_{\tilde{\mathbf{x}}} \left[\mathbf{s}_\theta(\tilde{\mathbf{x}})^T \nabla_{\tilde{\mathbf{x}}} \log p_\sigma(\tilde{\mathbf{x}}) \right] + C,$$

and analogously

$$\mathbb{E}_{\mathbf{x},\tilde{\mathbf{x}}}\left[\|\nabla_{\tilde{\mathbf{x}}}\log p(\tilde{\mathbf{x}}|\mathbf{x}) - \mathbf{s}_\theta(\tilde{\mathbf{x}})\|_2^2\right] = \mathbb{E}_{\tilde{\mathbf{x}}}\left[\|\mathbf{s}_\theta(\tilde{\mathbf{x}})\|_2^2\right] + \mathbb{E}_{\mathbf{x},\tilde{\mathbf{x}}}\left[\mathbf{s}_\theta(\tilde{\mathbf{x}})^T\nabla_{\tilde{\mathbf{x}}}\log p(\tilde{\mathbf{x}}|\mathbf{x})\right] + C.$$

We now show that the expectations over the cross terms above are equivalent, which proves the result. The first cross-term can be written as

$$\begin{aligned}
\mathbb{E}_{\tilde{\mathbf{x}}}\left[\mathbf{s}_\theta(\tilde{\mathbf{x}})^T\nabla_{\tilde{\mathbf{x}}}\log p_\sigma(\tilde{\mathbf{x}})\right] &= \int p_\sigma(\tilde{\mathbf{x}})\mathbf{s}_\theta(\tilde{\mathbf{x}})^T\nabla_{\tilde{\mathbf{x}}}\log p_\sigma(\tilde{\mathbf{x}})d\tilde{\mathbf{x}} \\
&= \int p_\sigma(\tilde{\mathbf{x}})\mathbf{s}_\theta(\tilde{\mathbf{x}})^T\frac{1}{p_\sigma(\tilde{\mathbf{x}})}\nabla_{\tilde{\mathbf{x}}}p_\sigma(\tilde{\mathbf{x}})d\tilde{\mathbf{x}} \\
&= \int \mathbf{s}_\theta(\tilde{\mathbf{x}})^T\nabla_{\tilde{\mathbf{x}}}\int p_\sigma(\tilde{\mathbf{x}}|\mathbf{x})p(\mathbf{x})d\mathbf{x}d\tilde{\mathbf{x}} \\
&= \int \mathbf{s}_\theta(\tilde{\mathbf{x}})^T\int p(\mathbf{x})p_\sigma(\tilde{\mathbf{x}}|\mathbf{x})\nabla_{\tilde{\mathbf{x}}}\log p_\sigma(\tilde{\mathbf{x}}|\mathbf{x})d\mathbf{x}d\tilde{\mathbf{x}} \\
&= \int\int p(\mathbf{x},\tilde{\mathbf{x}})\mathbf{s}_\theta(\tilde{\mathbf{x}})^T\nabla_{\tilde{\mathbf{x}}}\log p_\sigma(\tilde{\mathbf{x}}|\mathbf{x})d\mathbf{x}d\tilde{\mathbf{x}} \\
&= \mathbb{E}_{\mathbf{x},\tilde{\mathbf{x}}}\left[\mathbf{s}_\theta(\tilde{\mathbf{x}})^T\nabla_{\tilde{\mathbf{x}}}\log p_\sigma(\tilde{\mathbf{x}}|\mathbf{x})\right],
\end{aligned}$$

which concludes the proof of Equation (10.19).

10.2.3 Annealed Langevin dynamics

After training, samples can be produced with annealed Langevin dynamics [SE19; SE20] by running Langevin dynamics for $i = 1, \ldots, L$. Specifically, we start by initializing $\mathbf{x}_0 \sim \mathcal{N}(0, \mathbf{I})$, and start with noise scale $i = 1$ and step size $\alpha_i = \alpha\frac{\sigma_i^2}{\sigma_L^2}$ and run Langevin dynamics for T steps according to

$$\mathbf{x}_{t+1} = \mathbf{x}_t + \frac{\alpha_i}{2}\mathbf{s}(\mathbf{x}_t, \sigma_i) + \sqrt{\alpha_i}\epsilon_t, \quad \epsilon_t \sim \mathcal{N}(0, \mathbf{I}). \tag{10.20}$$

Then we initialize $\mathbf{x}_0$ with $\mathbf{x}_T$ and run Langevin dynamics with the next noise scale $i = 2$, and so on, until we ran Langevin dynamics for each noise scale.

10.3 Stochastic differential equations

In the first two sections, we introduced diffusion models as a discrete iterative process and with a score-based perspective. In this section we introduce a third perspective based on stochastic differential equations, which was popularized by Song et al. [Son+21].

Consider a diffusion process $\mathbf{x}(t)$ indexed by a continuous variable $t \in [0, T]$, where $\mathbf{x}(0) \sim p_0$, where p_0 is the data distribution and $\mathbf{x}(T) \sim p_T$, where p_T is the prior distribution. The data distribution can be modelled as the solution to a stochastic differential equation

$$d\mathbf{x} = \mathbf{f}(\mathbf{x}, t)dt + g(t)d\mathbf{w}.$$

Here, $\mathbf{f}(\cdot, t)$ is a vector-valued function called the drift coefficient which represents the deterministic part of the motion. The drift coefficient determines the evolution over time without the noise. The diffusion coefficient g is a scalar function that quantifies the intensity of the stochastic or random component of the process. Finally, $\mathbf{w}$ is Brownian motion.

Brownian motion is a random process where

(i) the increments $\mathbf{w}(t) - \mathbf{w}(s)$ over non-overlapping intervals are independent,
(ii) the increments are stationary, i.e., they only depend on the time interval $t - s$, and not on the starting time, and
(iii) the increments are Gaussian, i.e., for all t, s with $t > s$, we have

$$\mathbf{w}(t) - \mathbf{w}(s) \sim \mathcal{N}(0, (t - s)\mathbf{I}).$$

Let us start with an example of a familiar process that can be modelled with a stochastic differential equation.

Example: The stochastic gradient method. The stochastic gradient method iterations for minimizing a function f with stochastic gradient $G(\mathbf{x}) = \nabla f(\mathbf{x}) + \epsilon$ and step size β_i are given by

$$\mathbf{x}_{i+1} = \mathbf{x}_i - \beta_i \nabla f(\mathbf{x}_i) - \epsilon_i. \tag{10.21}$$

Here, ϵ is zero-mean random noise. The random variable $G(\mathbf{x})$ is called a stochastic gradient since by definition, in expectation over the noise ϵ, it is equal to the gradient of the function at $\mathbf{x}$, i.e., $\mathbb{E}[G(\mathbf{x})] = \nabla f(\mathbf{x})$. Note that typically for the stochastic gradient method, the step size also multiplies with the noise term, but for the transition to a stochastic differential equation, it is simpler to not have the noise ϵ_i multiplied with the step size.

Suppose the iterates $\mathbf{x}_i$ are discrete samples of a continuous time function $\mathbf{x}(t)$, i.e., $\mathbf{x}_i = \mathbf{x}(i/N)$, where we used the discretization scheme $\Delta t = 1/N$ and $t \in [0, 1/N, \ldots, (N-1)/N]$, and we let $\epsilon(i/N) = \epsilon_i$. With this, we can write the iterates (10.21) as

$$\mathbf{x}(t + \Delta t) = \mathbf{x}(t) - \beta(t)\Delta t \nabla f(\mathbf{x}(t)) + \boldsymbol{\epsilon}(t),$$

where we set $\beta_i = \beta(i/N)\Delta t$. Next, we define a random process $\mathbf{w}(t)$ so that the noise of the discretization corresponds to the increments of the continuous time process, i.e., $\boldsymbol{\epsilon}(t) = \mathbf{w}(t + \Delta t) - \mathbf{w}(t)$. This yields

$$\mathbf{x}(t + \Delta t) - \mathbf{x}(t) = -\beta(t)\Delta t \nabla f(\mathbf{x}(t)) + \mathbf{w}(t + \Delta t) - \mathbf{w}(t),$$

which in turn, for infinitesimal Δt, becomes

$$d\mathbf{x} = -\beta(t)\nabla f(\mathbf{x}(t))dt + d\mathbf{w}.$$

This has the form

$$d\mathbf{x} = \mathbf{f}(\mathbf{x}, t)dt + g(t)d\mathbf{w},$$

which is a generic form of an SDE with drift coefficient $\mathbf{f}(\mathbf{x}, t)$ and diffusion coefficient $g(t)$. For this example, the drift coefficient is $\mathbf{f}(\mathbf{x}, t) = -\beta(t)\nabla f(\mathbf{x}(t))$ and the diffusion coefficient is $g(t) = 1$. Thus, the deterministic part of the motion moves the motion in the direction of the negative gradient.

10.3.1 Forward and backward iterations

For diffusion models, the stochastic differential equation

$$d\mathbf{x} = \mathbf{f}(\mathbf{x}, t)dt + g(t)d\mathbf{w}$$

models the forward diffusion process, i.e., how to transition from the data distribution $\mathbf{x}(0) \sim p_0$ to the prior distribution $\mathbf{x}(T) \sim p_T$. The reverse of a diffusion process is also a diffusion process run backwards in time and is given according to Anderson [And82] by the following reverse-time SDE:

$$d\mathbf{x} = \left(f(\mathbf{x}, t) - g(t)^2 \nabla_{\mathbf{x}} \log p_t(\mathbf{x})\right) dt + g(t)d\bar{\mathbf{w}}, \tag{10.22}$$

where $\bar{\mathbf{w}}$ is Brownian motion where the time flows backwards from T to 0, and $\nabla_{\mathbf{x}} \log p_t(\mathbf{x})$ is the score function of the probability distribution of the state $\mathbf{x}$ at time t.

Example: Gaussian to a Gaussian with a stochastic differential equation. Let us consider a simple example. Suppose we have a data distribution p_0 that is Gaussian with mean μ_0 and variance σ_0^2, and a prior distribution that is also Gaussian, but has mean μ_T and variance $\sigma_T^2 \geq \sigma_0^2$. Consider a drift function that linearly interpolates between the two means over the time interval $[0, T]$, i.e., $f(x,t) = \frac{\mu_T-\mu_0}{T}$, and set the diffusion coefficient to the constant $g(t) = \sqrt{\frac{\sigma_T^2-\sigma_0^2}{T}}$. This yields the stochastic differential equation

$$dx = \frac{\mu_T - \mu_0}{T} dt + \sqrt{\frac{\sigma_T^2 - \sigma_0^2}{T}} dw. \tag{10.23}$$

By our definition of the drift function, the mean of $\mathbf{x}(T)$ is μ_T, as desired. To see this, note that since the drift is linear, the expected value at an intermediate time t is $\mu_0 + \frac{t}{T}(\mu_T - \mu_0)$, which becomes μ_T at $t = T$. The expected value at an intermediate time is the initial mean μ_0 plus the integral of the drift term, i.e., $\int_0^t \frac{\mu_T-\mu_0}{T} dt = t\frac{\mu_T-\mu_0}{T}$.

The variance only evolves due to the stochastic term, as the drift coefficient does not contribute to the variance. The variance at time t is equal to the initial variance, σ_0^2, plus the variance due to the diffusion term which yields:

$$\sigma_t^2 = \sigma_0^2 + \int_0^T \frac{\sigma_T^2 - \sigma_0^2}{T} ds = \sigma_0^2 + \frac{t}{T}(\sigma_T^2 - \sigma_0^2) = \sigma_T^2,$$

where we used the definition of the diffusion coefficient. Thus, the stochastic term accumulates variance over time.

See Figure 10.5 for a few example runs of the diffusion process modelled by the SDE (10.23).

Now let us derive the reverse-time SDE. The density $p_t(\mathbf{x})$ is Gaussian, i.e., $\mathcal{N}(\mu_t = \mu_0 + t\frac{\mu_T-\mu_0}{T}, \sigma_t^2 = \sigma_0^2 + \frac{t}{T}(\sigma_T^2 - \sigma_0^2))$, thus we can compute the reverse-time SDE as

$$\begin{aligned} dx &= \left(f(x,t) - g(t)^2 \nabla_x \log p_t(vx)\right) dt + g(t) d\bar{w} \\ &= \left(\frac{\mu_T - \mu_0}{T} - \frac{\sigma_T^2 - \sigma_0^2}{T} \frac{x - \mu_t}{\sigma_t^2}\right) dt + \sqrt{\frac{\sigma_T^2 - \sigma_0^2}{T}} d\bar{w}. \end{aligned}$$

Running time backwards from T to 0 yields again a Gaussian with mean μ_0 and variance σ_0^2.

10.3.2 Forward and backward iterations for DDPM

Recall that the iterates of the DDPM parameterization are (see Equation (10.1))

$$\mathbf{x}_i = \sqrt{1 - \beta_i}\mathbf{x}_{i-1} + \sqrt{\beta_i}\epsilon_i, \quad \epsilon_i \sim \mathcal{N}(0, \mathbf{I}), \tag{10.24}$$

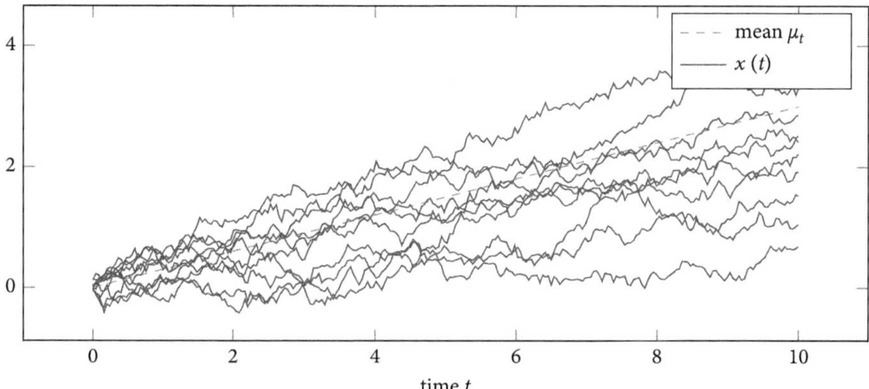

Figure 10.5 Example instances of the diffusion process modelled by the SDE (10.23) for $\mu_0 = 0, \sigma_0 = 0.1$ and $\mu_T = 3, \sigma_T = 1$.

where we used the iteration variable i (instead of t) to distinguish it from the continuous time variable t. Those iterates correspond to the following stochastic differential equation as shown by Song et al. [Son+21]:

$$d\mathbf{x} = -\frac{\beta(t)}{2}\mathbf{x}dt + \sqrt{\beta(t)}d\mathbf{w}. \tag{10.25}$$

To see this, one can proceed with defining a step size $\Delta t = 1/N$, setting $\beta_t = \beta(t/N)\Delta t$ for $t = 0, \ldots, (N-1)/N$ and proceeding with analogous calculations as for the discretization of the stochastic gradient method above starting from Equation (10.24), see the paragraph below for the details.

Since the DDPM forward iterates can be written as a stochastic differential equation, the DDPM estimate can be obtained by solving the stochastic differential equation.

For DDPM, the reverse-time SDE (10.22) becomes (with $f(\mathbf{x}, t) = -\frac{\beta(t)}{2}\mathbf{x}$ and $g(t) = \sqrt{\beta(t)}$)

$$d\mathbf{x} = -\beta(t)\left(\frac{1}{2}\mathbf{x} + \nabla_{\mathbf{x}} \log p_t(\mathbf{x})\right)dt + \sqrt{\beta(t)}d\bar{\mathbf{w}}. \tag{10.26}$$

With our discretization scheme ($\Delta t = 1/N$, $\mathbf{x}(i/N) = \mathbf{x}_i$ and $\beta(i/N)\Delta t = \beta_i$), the discrete analogue of the reverse SDE is (see below for the derivation and the approximations involved):

$$\mathbf{x}_{i-1} = \frac{1}{\sqrt{1-\beta_i}}\left(\mathbf{x}_i + \frac{\beta_i}{2}\nabla_{\mathbf{x}_i} \log p_i(\mathbf{x}_i)\right) + \sqrt{\beta_i}\epsilon_i. \tag{10.27}$$

The sampling process described by Equation (10.27) was termed ancestral sampling by Song et al. [Son+21], where ancestral refers to the process of generating samples step by step. Running the discrete analogue of the reverse SDE, i.e., performing ancestral sampling enables us to obtain samples from the diffusion model. This sampling process is interesting as it involves the score function.

Derivation of Equation (10.25): Similar as in our discussion of the stochastic gradient method, we use the step size $\Delta t = 1/N$ and define the function $\beta(t)$ so that $\beta_i = \beta(i/N)\Delta t$, and set $\mathbf{x}_i = \mathbf{x}(i/N)$. Moreover, let $\mathbf{w}(t)$ be Brownian motion with (scaled) increments in the time interval Δt $\frac{\mathbf{w}(t)-\mathbf{w}(t-\Delta t))}{\sqrt{\Delta t}} = \epsilon(t = i/N) = \epsilon_i$. Note that the scaled increments are standard normal distributed, i.e., $\epsilon_i \sim \mathcal{N}(0, \mathbf{I})$ by definition of Brownian motion. With this notation, the DDPM iterates (10.24) become

$$\mathbf{x}(t) = \sqrt{1 - \beta(t)\Delta t}\mathbf{x}(t - \Delta t) + \sqrt{\beta(t)\Delta t}\frac{\mathbf{w}(t) - \mathbf{w}(t - \Delta t)}{\sqrt{\Delta t}}.$$

Using the first order Taylor expansion of $\sqrt{1-\beta}$ around $\beta = 0$, we have that for small β, $\sqrt{1-\beta} = 1 - \frac{\beta}{2}$. With this approximation, and re-arranging terms, we obtain

$$\mathbf{x}(t) - \mathbf{x}(t - \Delta t) \approx -\frac{\beta(t)\Delta t}{2}\mathbf{x}(t - \Delta t) + \sqrt{\beta(t)}(\mathbf{w}(t) - \mathbf{w}(t - \Delta t)).$$

Thus, as Δt vanishes, this becomes the stochastic differential equation

$$d\mathbf{x} = -\frac{\beta(t)}{2}\mathbf{x}dt + \sqrt{\beta(t)}d\mathbf{w},$$

as claimed.

Derivation of Equation (10.27): Define the (scaled) increments of the Brownian motions in the time interval Δt as $\epsilon_i = \epsilon(t = i/N) = \frac{\mathbf{w}(t)-\mathbf{w}(t-\Delta t))}{\sqrt{\Delta t}}$ and note that the increments are standard normal distributed, i.e., $\epsilon_i \sim \mathcal{N}(0, \mathbf{I})$, because the increments of the Brownian motion, $\mathbf{w}(t) - \mathbf{w}(t - \Delta t)$, over any time interval Δt are normally distributed with mean 0 and co-variance $\Delta t\mathbf{I}$. With this notation,

$$\mathbf{x}(t) - \mathbf{x}(t - \Delta t) = -\beta(t)\left(\frac{1}{2}\mathbf{x}(t) + \nabla_{\mathbf{x}} \log p_t(\mathbf{x}(t))\right)\Delta t - \sqrt{\beta(t)\Delta t}\epsilon(t).$$

Re-arranging terms yields

$$\begin{aligned}
\mathbf{x}(t - \Delta t) &= \mathbf{x}(t) + \beta(t)\left(\frac{1}{2}\mathbf{x}(t) + \nabla_{\mathbf{x}} \log p_t(\mathbf{x}(t))\right)\Delta t + \sqrt{\beta(t)\Delta t}\epsilon(t) \\
&= \mathbf{x}(t)\left(1 + \frac{\beta(t)\Delta t}{2}\right) + \nabla_{\mathbf{x}} \log p_t(\mathbf{x}(t))\beta(t)\Delta t + \sqrt{\beta(t)\Delta t}\epsilon(t) \\
&\approx \left(1 + \frac{\beta(t)\Delta t}{2}\right)(\mathbf{x}(t) + \nabla_{\mathbf{x}} \log p_t(\mathbf{x}(t))\beta(t)\Delta t) + \sqrt{\beta(t)\Delta t}\epsilon(t),
\end{aligned}$$

where the approximation holds when assuming that $\beta(t)\Delta t \ll 1$. With our discretization scheme $\Delta t = 1/N$, $\mathbf{x}(i/N) = \mathbf{x}_i$, $\beta(i/N)\Delta t = \beta_i$, and $\epsilon_i = \epsilon(i/N)$, this becomes

$$\begin{aligned}\mathbf{x}_{i-1} &\approx \left(1 + \frac{1}{2}\beta_i\right)\left(\mathbf{x}_i + \frac{\beta_i}{2}\nabla_{\mathbf{x}_i} \log p_i(\mathbf{x}_i)\right) + \sqrt{\beta_i}\epsilon_i \\ &\approx \frac{1}{\sqrt{1-\beta_i}}\left(\mathbf{x}_i + \frac{\beta_i}{2}\nabla_{\mathbf{x}_i} \log p_i(\mathbf{x}_i)\right) + \sqrt{\beta_i}\epsilon_i.\end{aligned}$$

Here, the approximation $1/\sqrt{1-\beta} \approx 1 + \beta/2$ holds for small β and follows by the Taylor series expansion $1/\sqrt{1-\beta} = 1 + 1/2\beta + 3/8\beta^2 + \ldots$ around $\beta = 0$. This concludes the derivation of Equation (10.27).

10.4 Solving inverse problems with diffusion and score-based models

Consider a standard linear inverse problem, where we are given a noisy measurement $\mathbf{y} = \mathbf{A}\mathbf{x} + \mathbf{e} \in \mathbb{R}^m$ of an image $\mathbf{x} \in \mathbb{R}^n$ and our goal is to estimate the image. Here, $\mathbf{A}$ is a linear forward model and $\mathbf{e} \sim \mathcal{N}(0, \sigma^2\mathbf{I})$ is Gaussian noise.

Diffusion models are very successful for image and video generation since they are able to provide powerful image priors. Those priors can be leveraged to solve inverse problems by conditionally sampling from the distribution of the images given the measurement.

Exactly sampling from the conditional distribution with density $p(\mathbf{x}|\mathbf{y})$ with diffusion and score-based models is intractable as we will see later. Thus, current methods for conditionally sampling based on diffusion models leverage approximations. We discuss three approaches for solving a linear inverse problem with diffusion models.

- **Reconstruction with Langevin dynamics.** This approach is based on using the score function to guide each diffusion sampling iteration via Langevin sampling.
- **Posterior sampling.** Posterior sampling is based on estimating the conditional score function $\nabla_{\mathbf{x}_t} \log p(\mathbf{x}_t|\mathbf{y})$ and using the score function in ancestral sampling as discussed above.
- **Variational reconstruction.** The third method is based on taking a variational approach, i.e., formulating reconstruction as an optimization problem involving the diffusion model as a regularizer.

10.4.1 Reconstruction with Langevin dynamics

We can reconstruct by approximately sampling with the conditional density $p(\mathbf{x}|\mathbf{y})$. Langevin dynamics enables sampling from the conditional distribution according to the Langevin dynamics iterations (10.13), all we need for this is the score function of the conditional distribution, i.e., $\nabla_{\mathbf{x}_t} \log p(\mathbf{x}_t|\mathbf{y})$. Bayes' rule for the conditional density states that

$$p(\mathbf{x}_t|\mathbf{y}) = \frac{p(\mathbf{y}|\mathbf{x}_t)p(\mathbf{x}_t)}{p(\mathbf{y})}.$$

Thus the score function of the conditional distribution is

$$\nabla_{\mathbf{x}_t} \log p(\mathbf{x}_t|\mathbf{y}) = \nabla_{\mathbf{x}_t} \log p(\mathbf{y}|\mathbf{x}_t) + \nabla_{\mathbf{x}_t} \log p(\mathbf{x}_t).$$

For our setup where the measurement noise $\mathbf{e}$ is Gaussian, the conditional distribution $p(\mathbf{y}|\mathbf{x}_0)$ is Gaussian with mean $\mathbf{A}\mathbf{x}_0$ and co-variance matrix $\sigma^2\mathbf{I}$. Thus, the score function of the conditional distribution with density function $p(\mathbf{y}|\mathbf{x}_0)$ is

$$\begin{aligned}\nabla_{\mathbf{x}_0} \log p(\mathbf{y}|\mathbf{x}_0) &= -\nabla_{\mathbf{x}_0} \frac{1}{2\sigma^2} \|\mathbf{A}\mathbf{x}_0 - \mathbf{y}\|_2^2 \\ &= -\frac{1}{\sigma^2}\mathbf{A}^T(\mathbf{A}\mathbf{x}_0 - \mathbf{y}).\end{aligned}$$

For the other noise levels $t \neq 0$, the score is not equal to the form above but Jalal et al. [Jal+21] proposed to approximate it with

$$\nabla_{\mathbf{x}_t} \log p(\mathbf{y}|\mathbf{x}_t) \approx -\frac{1}{\sigma^2 + \gamma_t^2}\mathbf{A}^T(\mathbf{A}\mathbf{x}_0 - \mathbf{y}),$$

where $\gamma_t, t = 1, 2, \ldots$ are hyperparameters. Now suppose we trained a noise-conditional score model $\mathbf{s}_\theta(\mathbf{x}, \sigma)$ as discussed above. Given this noise-conditional score model, we can run Langevin dynamics. As before, we start by initializing $\mathbf{x}_0 \sim \mathcal{N}(0, \mathbf{I})$, and start with noise scale $i = 1$ and step size $\alpha_i = \alpha\frac{\sigma_i^2}{\sigma_L^2}$ and run Langevin dynamics for T steps according to

$$\mathbf{x}_{t+1} = \mathbf{x}_t + \frac{\alpha_i}{2}\left(-\frac{1}{\sigma^2 + \gamma_i^2}\mathbf{A}^T(\mathbf{A}\mathbf{x} - \mathbf{y}) + \mathbf{s}_\theta(\mathbf{x}_t, \sigma_i)\right) + \sqrt{\alpha_i}\epsilon_t, \quad \epsilon_t \sim \mathcal{N}(0, \mathbf{I}).$$

Then we initialize $\mathbf{x}_0$ with $\mathbf{x}_T$ and run Langevin dynamics with the next noise scale $i = 2$, and so on, until we ran Langevin dynamics for each noise scale.

This version of annealed Langevin dynamics for medical image reconstruction in the context of MRI has been proposed by Jalal et al. [Jal+21] and performs relatively well.

10.4.2 Posterior sampling

We now discuss an approach for solving an inverse problem with a diffusion model based on estimating the posterior distribution of the iterate $\mathbf{x}_t$ given the measurement $\mathbf{y}$. This approach requires estimation of the score function of the conditional posterior, $\nabla_{\mathbf{x}_t} \log p(\mathbf{x}_t|\mathbf{y})$, at each step t of the reverse diffusion process. With the estimates of the score function $\nabla_{\mathbf{x}_t} \log p(\mathbf{x}_t|\mathbf{y})$ we can simply solve the reverse stochastic differential equation (10.26), for example by performing ancestral sampling according to Equation (10.27), by using the conditional score function $\nabla_{\mathbf{x}_t} \log p(\mathbf{x}_t|\mathbf{y})$ instead of the unconditional score function $\nabla_{\mathbf{x}_t} \log p(\mathbf{x}_t)$. We now explain how the conditional score function $\nabla_{\mathbf{x}_t} \log p(\mathbf{x}_t|\mathbf{y})$ can be estimated.

With Bayes' theorem, the score function of the conditional distribution can be written as

$$\begin{aligned}\nabla_{\mathbf{x}_t} \log p(\mathbf{x}_t|\mathbf{y}) &= \nabla_{\mathbf{x}_t} \log \frac{p(\mathbf{y}|\mathbf{x}_t)p(\mathbf{x}_t)}{p(\mathbf{y})} \\ &= \nabla_{\mathbf{x}_t} \log p(\mathbf{y}|\mathbf{x}_t) + \nabla_{\mathbf{x}_t} \log p(\mathbf{x}_t).\end{aligned}$$

We are given an estimate of the score function $\nabla_{\mathbf{x}_t} \log p(\mathbf{x}_t)$ from the pre-trained diffusion model. The difficulty of evaluating the right-hand side above is estimating the score function of the conditional distribution $\nabla_{\mathbf{x}_t} \log p(\mathbf{y}|\mathbf{x}_t)$. Song et al. [Son+22] and Chung et al. [Chu+23a] estimate this term as follows. First, note that

$$p(\mathbf{y}|\mathbf{x}_t) = \int p(\mathbf{y}|\mathbf{x}_0)p(\mathbf{x}_0|\mathbf{x}_t)d\mathbf{x}_0. \tag{10.28}$$

In this equation, the conditional density $p(\mathbf{y}|\mathbf{x}_0)$ is Gaussian with mean $\mathbf{A}\mathbf{x}_0$ and co-variance matrix $\sigma^2\mathbf{I}$. The conditional density $p(\mathbf{x}_0|\mathbf{x}_t)$ can be approximated as follows.

Song et al. [Son+22] approximate the density $p(\mathbf{x}_0|\mathbf{x}_t)$ with a Gaussian $\mathcal{N}(\hat{\mathbf{x}}_t, \varsigma_t^2\mathbf{I})$ with mean

$$\hat{\mathbf{x}}_t = \mathbb{E}\left[\mathbf{x}_0|\mathbf{x}_t\right] = \mathbf{x}_t + \sigma_t^2 \nabla_{\mathbf{x}_t} \log p(\mathbf{x}_t),$$

where the second equality holds by Tweedie's formula. Since the measurement model is Gaussian, with the Gaussian approximation, we obtain

$$p(\mathbf{y}|\mathbf{x}_t) \approx \mathcal{N}(\mathbf{A}\hat{\mathbf{x}}_t, \varsigma_t^2\mathbf{A}\mathbf{A}^T + \sigma^2\mathbf{I}).$$

Here, we computed the integral in Equation (10.28) by using that the convolution of two Gaussian distributions $p(\mathbf{y}|\mathbf{x}_0)$ and $p(\mathbf{x}_0|\mathbf{x}_t)$ is again a Gaussian distribution.

With this, the score function can be approximated as

$$\nabla_{\mathbf{x}_t} \log p(\mathbf{y}|\mathbf{x}_t) \approx \nabla_{\mathbf{x}_t}(\mathbf{y} - \mathbf{A}\hat{\mathbf{x}}_t(\mathbf{x}_t))^T(\varsigma_t^2 \mathbf{A}\mathbf{A}^T + \sigma^2 \mathbf{I})^{-1}(\mathbf{y} - \mathbf{A}\hat{\mathbf{x}}_t(\mathbf{x}_t)),$$

where the gradient can be computed with automatic differentiation. Song et al. [Son+22] used this estimate to solve the reverse SDE (10.26), and achieved good results for accelerated MRI.

Chung et al. [Chu+23a] proposed a very related method, where the score is approximated slightly differently by

$$\nabla_{\mathbf{x}_t} \log p(\mathbf{y}|\mathbf{x}_t) \approx -\frac{1}{\sigma^2} \nabla_{\mathbf{x}_t} \|\mathbf{y} - \mathbf{A}\hat{\mathbf{x}}_0(\mathbf{x}_t)\|_2^2, \tag{10.29}$$

where

$$\hat{\mathbf{x}}_0 = \mathbb{E}\left[\mathbf{x}_0|\mathbf{x}_t\right] = \frac{1}{\sqrt{1-\sigma_t^2}} \left(\mathbf{x}_t + \sigma_t^2 \nabla_{\mathbf{x}_t} \log p(\mathbf{x}_t)\right).$$

Chung et al. [Chu+23a] used this estimate to solve the reverse SDE (10.26) via ancestral sampling, and achieved good results for accelerated MRI. Figure 10.6 shows an example reconstruction with this method.

10.4.3 Variational reconstruction

Finally, we discuss a variational approach to solve inverse problems with diffusion models by Mardani et al. [Mar+23]. The approach is based on minimizing the KL divergence between a variational distribution q and the conditional distribution p of the signal $\mathbf{x}_0$ conditioned on the measurement $\mathbf{y}$. We take the variational distribution q as a Gaussian with co-variance matrix $\sigma_q^2 \mathbf{I}$ and mean $\boldsymbol{\mu}$. Minimizing the KL divergence over the variational distribution amounts to solving the optimization problem

$$\min_{\boldsymbol{\mu}, \sigma_q} D_{KL}(q(\mathbf{x}_0|\mathbf{y}) \parallel p(\mathbf{x}_0|\mathbf{y})). \tag{10.30}$$

The KL divergence can be written as

$$\begin{aligned} D_{KL}(q(\mathbf{x}_0|\mathbf{y}) \parallel p(\mathbf{x}_0|\mathbf{y})) &= \mathbb{E}_{q(\mathbf{x}_0|\mathbf{y})}\left[\log \frac{q(\mathbf{x}_0|\mathbf{y})}{p(\mathbf{x}_0|\mathbf{y})}\right] \\ &= \mathbb{E}_{q(\mathbf{x}_0|\mathbf{y})}\left[\log \frac{q(\mathbf{x}_0|\mathbf{y})p(\mathbf{y})}{p(\mathbf{y}|\mathbf{x}_0)p(\mathbf{x}_0)}\right] \\ &= -\mathbb{E}_{q(\mathbf{x}_0|\mathbf{y})}\left[\log p(\mathbf{y}|\mathbf{x}_0)\right] + \mathbb{E}_{q(\mathbf{x}_0|\mathbf{y})}\left[\log \frac{q(\mathbf{x}_0|\mathbf{y})}{p(\mathbf{x}_0)}\right] + \log p(\mathbf{y}) \\ &= -\mathbb{E}_{q(\mathbf{x}_0|\mathbf{y})}\left[\log p(\mathbf{y}|\mathbf{x}_0)\right] + D_{KL}(q(\mathbf{x}_0|\mathbf{y}) \parallel p(\mathbf{x}_0)) + \log p(\mathbf{y}), \end{aligned} \tag{10.31}$$

where the first equation is by Bayes' theorem. The last term above (i.e., $\log p(\mathbf{y})$) does not depend on the distribution q and is thus irrelevant for the minimization problem (10.30).

Regarding the first term in Equation (10.31) above, recall that by our measurement model, the density $p(\mathbf{y}|\mathbf{x}_0)$ is Gaussian with mean $\mathbf{A}\mathbf{x}_0$ and co-variance matrix $\sigma^2\mathbf{I}$. Thus, the term is given by the reconstruction loss

$$-\mathbb{E}_{q(\mathbf{x}_0|\mathbf{y})}\left[\log p(\mathbf{y}|\mathbf{x}_0)\right] = \frac{1}{2\sigma^2}\mathbb{E}_{q(\mathbf{x}_0|\mathbf{y})}\left[\|\mathbf{A}\mathbf{x}_0 - \mathbf{y}\|_2^2\right]$$

which can be approximated by

$$-\mathbb{E}_{q(\mathbf{x}_0|\mathbf{y})}\left[p(\mathbf{y}|\mathbf{x}_0)\right] = \frac{1}{2\sigma^2}\|\mathbf{A}\boldsymbol{\mu} - \mathbf{y}\|_2^2$$

if the variance of the variational distribution, σ_q^2, is very small. As shown by Mardani et al. [Mar+23], for small values of the variance σ_q^2, the second term in Equation (10.31) can be approximated by

$$R(\boldsymbol{\mu}) = \mathbb{E}_{t,\epsilon}\left[w_t\|\epsilon_\theta(\mathbf{x}_t, t) - \epsilon\|_2^2\right], \tag{10.32}$$

where w_t is a weighting term and $\mathbf{x}_t = \sqrt{1-\sigma_t^2}\boldsymbol{\mu} + \sigma_t\epsilon$. Using those approximations in Equation (10.31) yields the variational objective

$$\mathcal{L}(\boldsymbol{\mu}) = \frac{1}{2\sigma^2}\|\mathbf{A}\boldsymbol{\mu} - \mathbf{y}\|_2^2 + \mathbb{E}_{t,\epsilon}\left[w_t\|\epsilon_\theta(\mathbf{x}_t, t) - \epsilon\|_2^2\right], \tag{10.33}$$

where $\epsilon_\theta(\mathbf{x}_t, t)$ is the neural network that predicts the noise ϵ which when added to the image yields the state $\mathbf{x}_{t-1}$ (see Equation (10.3)). To estimate the signal $\mathbf{x}_0$, the variational objective can be optimized with first order optimization methods, such as gradient descent or the stochastic gradient method. The derivative is

$$\nabla_{\boldsymbol{\mu}}\mathcal{L}(\boldsymbol{\mu}) = \frac{1}{\sigma^2}\mathbf{A}^T(\mathbf{A}\boldsymbol{\mu} - \mathbf{y}) + \mathbb{E}_{t,\epsilon}\left[w_t\sqrt{1-\sigma_t^2}(\epsilon_\theta(\mathbf{x}_t, t) - \epsilon)^T\mathbf{J}\right], \tag{10.34}$$

where $\mathbf{J}$ is the Jacobian matrix of the function $\epsilon_\theta(\mathbf{x}_t, t)$. This derivative can be computed using automatic differentiation. However computing the derivative can be very costly because of the size of the Jacobian. Therefore, a variety of works [Mar+23; Poo+23] simply approximate the Jacobian with the identity matrix which yields good results.

Figure 10.6 shows an example reconstruction with this variational approach.

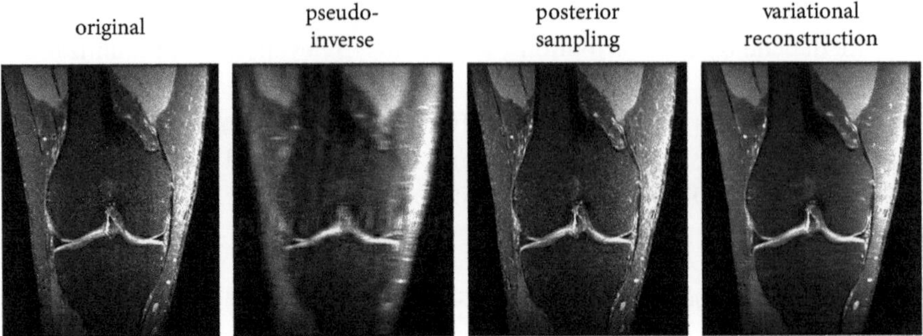

Figure 10.6 Reconstructions with the pseudo-inverse as a reference, posterior sampling, and variational reconstruction for 4-times accelerated single-coil MRI. The diffusion model is trained on slices extracted from 15 volumes from the Stanford Fullysampled 3D FSE Knees dataset by minimizing the DDPM objective (10.4). Posterior sampling is implemented with the approximation in Equation (10.29). It can be seen that the variational approach generates a smoother reconstruction than the ones generated by the posterior sampling-based approach, but is more accurate according to the image reconstruction metrics PSNR and SSIM. This is consistent with the results on a larger test set.

10.5 Chapter notes

In this chapter, we discussed diffusion models and how to use them for image reconstruction. We focused on 2D reconstruction. However, a particular advantage of diffusion model-based reconstruction and more broadly, reconstruction based on generative models, is that that those models provide powerful image models that can be used for a variety of complicated imaging tasks such as 3D and 4D imaging that are out of reach or very challenging for current supervised methods.

For example, 3D reconstruction is challenging for computational and for data reasons. Suppose we were to train a network for 3D reconstruction end-to-end. Working with 3D volumes is significantly more expensive computationally, and requires a lot of memory. Perhaps more importantly, it is very difficult to obtain high-quality 3D data for medical imaging such as MRI or CT. For such a challenging scenario, however, 2D diffusion priors can be leveraged. Specifically, we can train a 2D diffusion model on different 2D slices and use the 2D diffusion model as prior in the posterior sampling-based or in the variational reconstruction approach discussed in this chapter.

Moreover, reconstruction with generative and diffusion-based models enables to naturally quantify reconstruction uncertainty, since the models provide a prior over the image distribution.

11

Signal reconstruction with un-trained neural networks

The majority of deep learning-based signal reconstruction methods—and all we have seen so far—rely on training on a set of example images. In this chapter, we consider a class of un-trained methods that reconstruct an image based on a single measurement and do not rely on training data, apart from potential hyper-parameter tuning.

We focus on an un-trained method that is based on imposing a convolutional un-trained neural network as image prior by fitting the network with gradient descent to data at inference time. This idea has been introduced by Ulyanov et al. [UVL18] as the deep image prior (DIP). Ulyanov et al. [UVL18] proposed to fit a standard convolutional autoencoder (the popular U-net [RFB15]) to a single noisy image via gradient descent, and regularized by early stopping.

Un-trained networks have emerged as a highly successful alternative to data-driven methods, and yield excellent performance for a variety of problems, including denoising [UVL18; HH19] and compressive sensing [Yoo+21; VV+18; JH19; Bos+20; Wan+20; HA20].

Despite being neural network-based, this method is conceptually related to classical sparsity-based methods, in that it relies on imposing a hand-crafted signal prior in the form of a neural network. Un-trained networks are quite effective: They can provide significant improvements over sparsity-based methods. For example, they give better image quality for accelerated magnetic resonance imaging [DH21] than ℓ_1-regularized least-squares.

In the literature, un-trained methods are also sometimes discussed as self-supervised. We call this class of methods un-trained to emphasize that there is no reliance on training data unlike the self-supervised methods discussed in Chapter 8. Un-trained methods are also sometimes called zero-shot methods which is fitting as it emphasizes that there is no reliance on training data. However, zero-shot methods broadly refer to techniques where a model handles tasks it has not explicitly seen during training, and such methods are often trained on some data.

Deep Learning for Computational Imaging. Reinhard Heckel, Oxford University Press. © Reinhard Heckel (2025).
DOI: 10.1093/oso/9780198947172.003.0011

11.1 Overview of the method

We consider the problem of reconstructing a signal $\mathbf{x} \in \mathbb{R}^n$ from a noisy linear measurement $\mathbf{y} = \mathbf{A}\mathbf{x} + \mathbf{e} \in \mathbb{R}^m$.

Let $G: \mathbb{R}^p \to \mathbb{R}^n$ be a neural network with p parameters. In contrast to the previous section on signal reconstruction with a generative prior, where we viewed the network as a function of the input, here we view the network G as a function of the network weights $\boldsymbol{\theta} \in \mathbb{R}^p$. The input to the network, $\mathbf{z}$, is typically chosen at random and then fixed during updating the model's parameters. The function G is the un-trained neural network, and the fixed input $\mathbf{z}$ is considered part of the network.

The architecture of the network is critical, and is discussed in more detail later. For now, we note that a good choice for image reconstruction is a simple five-layer convolutional network, which is depicted in Figure 11.1.

We reconstruct an image by first minimizing the least-squares loss

$$\mathcal{L}(\boldsymbol{\theta}) = \|\mathbf{A}G(\boldsymbol{\theta}) - \mathbf{y}\|_2^2 \tag{11.1}$$

with an optimization procedure, typically gradient descent or Adam, which start from a random initialization of the network weights, $\boldsymbol{\theta}_0$, and then update the model parameters iteratively by performing gradient steps. This optimization procedure, possibly early-stopped for regularization, yields the estimate $\hat{\boldsymbol{\theta}}$, from which we estimate the unknown signal as $\hat{\mathbf{x}} = G(\hat{\boldsymbol{\theta}})$.

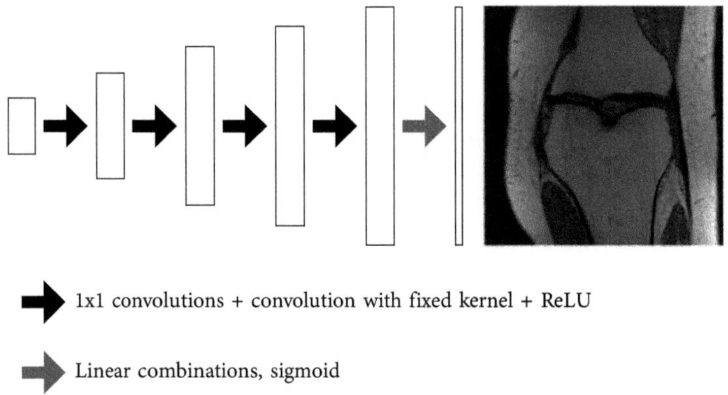

Figure 11.1 An illustration of the deep decoder, a five-layer un-trained convolutional neural network. The network performs 1x1 convolutions (i.e., linear combinations of channels) followed by convolutions with a fixed kernel to map one volume to another. Convolution with a fixed kernel often includes an upsampling operation, as displayed here.

This general approach is based on the empirical observation that un-trained convolutional networks fit a single natural image significantly faster than random noise when optimized with gradient descent, as observed by Ulyanov et al. [UVL18]. For the method to work well, a good choice of network architecture, optimization procedure, and regularization (e.g., early stopping) is important.

Below, we discuss a result showing that an un-trained (convolutional) neural network can provably denoise a smooth signal. An un-trained convolutional neural network can also provably estimate a smooth signal well from few random measurements [HS20a].

Deep image prior. Ulyanov et al. [UVL18] first observed that using a standard convolutional autoencoder (the popular U-net [RFB15]) as a generator network, and regularizing with early stopping, enables excellent denoising performance. This method has been termed *deep image prior.*

Deep decoder. Many elements of the U-net turn out to be irrelevant to the performance of the deep image prior. The paper [HH19] proposed a simpler network architecture, termed the deep decoder, shown in Figure 11.1. This network can be seen as retaining only the most relevant components of a convolutional autoencoder architecture to function as an image prior, and can be obtained from a standard convolutional autoencoder by removing the encoder, the skip-connections, and the trainable convolutional filters of spatial extent larger than one.

To understand un-trained networks better, in the following two sections, we discuss signal reconstruction with under- and over-parameterized un-trained networks. We say that a network is under-parameterized if it has fewer parameters (i.e., weights) than the network's output (i.e., the number of pixels of the image), and we say it is over-parameterized if the network has more parameters.

Detailed description of the deep decoder network architecture

A convolutional network $G: \mathbb{R}^p \to \mathbb{R}^n$ is a network that takes an input tensor and maps the input tensor to a signal through convolutional operations and applications of non-linearities. Here, we describe the deep decoder, a convolutional network, that uses convolutions with fixed (not trainable) filters only, which are important for a convolutional generator to function as an image model.

Let $\mathbf{B}_0 \in \mathbb{R}^{n_0 \times k}$ the input volume. The input volume is chosen randomly from a Gaussian distribution and is fixed. The channels in the $(i + 1)$-st layer are given by

$$\mathbf{B}_{i+1} = \text{cn}(\text{relu}(\mathbf{U}_i \mathbf{B}_i \boldsymbol{\theta}_i)), \quad i = 0, \ldots, d - 1,$$

and finally, the output of the d-layer network is formed as

$$\mathbf{x} = \mathbf{B}_d \boldsymbol{\theta}_{d+1} \in \mathbb{R}^{n \times k_{\text{out}}},$$

with $k_{\text{out}} = 1$ for a greyscale image, and $k_{\text{out}} = 3$ for an image with three colour channels. Here, n_i is an integer if the network generates a one-dimensional signal, and $n_i = w_i \times h_i$, if the network generates an image of width w and height h, which is a two-dimensional signal. The coefficient matrices $\boldsymbol{\theta}_i \in \mathbb{R}^{k \times k}$ and $\boldsymbol{\theta}_{d+1} \in \mathbb{R}^{k \times k_{out}}$ contain the weights of the network, and we denote with $\boldsymbol{\theta}$ the set of all those weight parameters. The number of channels, k, determines the number of weight parameters of the network, which is $dk^2 + k_{\text{out}}k$; choosing k larger increases the number of parameters of the channels. A typical value for the depth is $d = 5$.

Each column of the tensor $\mathbf{B}_i\boldsymbol{\theta}_i \in \mathbb{R}^{n_i \times k}$ is formed by taking linear combinations of the channels of the tensor $\mathbf{B}_i$ in a way that is consistent across all pixels. In deep learning language, this is often called a 1×1 because it is a degenerate convolution with kernel 1×1 for images.

The operator $\mathbf{U}_i \in \mathbb{R}^{n_{i+1} \times n_i}$ is a transposed convolution with a fixed convolutional kernel which is chosen as a triangular kernel. Thus, the operation performs bi-linear 2x upsampling.

Finally, $\text{cn}(\cdot)$ performs the channel normalization operation which normalizes each channel individually to zero-mean and unit variance and can be viewed as a special case of the popular batch normalization operation [IS15]. This is the least important element of the architecture for the conceptual understanding of the architecture, and can be viewed as a tool for enabling efficient optimization of the network, as discussed previously.

11.2 Signal reconstruction with an under-parameterized un-trained network

The convolutional network G introduced in the previous section maps a parameter vector $\boldsymbol{\theta} \in \mathbb{R}^p$ to an image. The network architecture enables to represent an image concisely, i.e., it can approximate a high-dimensional natural image well with much fewer parameters than the number of pixels of the image, as demonstrated next.

Figure 11.2 (from [HH19]) compares the performance of representing natural images with the convolutional network G to representing the same images with a sparse wavelet representation. The comparison to sparse wavelet representations is a natural choice since sparse wavelet representations are known to efficiently compress images. This characteristic underlies their use within the widely used lossy JPEG 2000 image format. The figure shows that a typical natural (ImageNet-) image can be approximated well with about 10–30 times fewer coefficients than the numbers of pixels (i.e., $q = n/30$ to $q = n/10$) with the deep decoder architecture introduced above. The figure also shows that the deep decoder architecture enables representations that are similar concise as sparse wavelet representations.

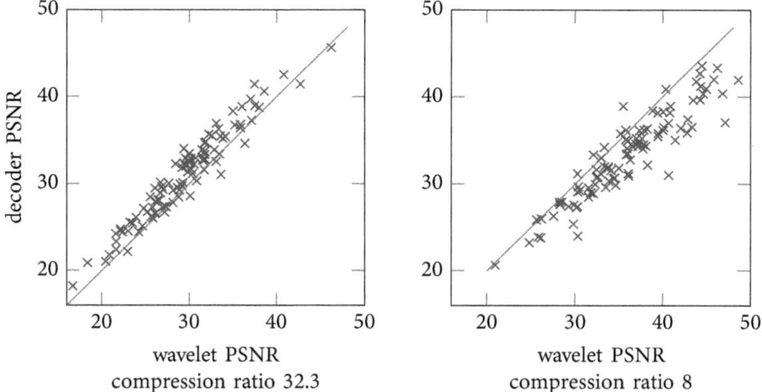

Figure 11.2 A convolutional network enables representing a signal concisely, on par with a sparse wavelet-based signal representation. The crosses on the left depict the PSNRs for 100 randomly chosen ImageNet-images represented with few wavelet coefficients and with a deep decoder with an equal number of parameters. A cross above the red line means the corresponding image has a smaller representation error when represented with the deep decoder. It can be seen that for a moderate to large compression ratio (32.3), the un-trained network yields higher quality for most images, while for a small compression ratio (8), the wavelet representation yields higher quality for most images.

Being able to represent an image concisely enables signal reconstruction via solving the optimization problem (11.1). To see this, consider a simple denoising problem, where the forward operator is the identity ($\mathbf{A} = \mathbf{I}$), and where the goal is to estimate a signal $\mathbf{x}$ from a noisy observation $\mathbf{y} = \mathbf{x} + \mathbf{e}$. Since the network is a good signal model, it can represent the signal part of the measurement $\mathbf{y}$ well, and since it is under-parameterized it can only represent a small proportion of random noise well. Specifically, an n-dimensional Gaussian noise vector requires roughly p parameters to represent a fraction of $\frac{p}{n}$ of its energy. For the two-layer deep decoder, this is formalized in the following proposition.

Proposition 11. *[HH19, Prop. 1] Consider a two-layer deep decoder with p parameters and arbitrary upsampling and input matrices. Let* $\mathbf{e}$ *be zero-mean Gaussian noise with identity covariance matrix. Then, with high probability,*

$$\min_{\boldsymbol{\theta}} \|G(\boldsymbol{\theta}) - \mathbf{e}\|_2^2 \geq \|\mathbf{e}\|_2^2 \left(1 - c\frac{p \log n}{n}\right), \tag{11.2}$$

where c is a numerical constant.

Proposition 11 reveals that when fitting an under-parameterized deep decoder to a noisy image (by minimizing the loss (11.1)), we expect to fit only a small amount of noise, thus enabling denoising. The number of network parameters, p, trades off how well the network fits the underlying signal (larger is better) and how much noise it fits (smaller is better).

Beyond denoising, similar ideas can be applied to proving reconstruction results for compressive sensing. Specifically, the proof of Proposition 11 establishes that any signal generated by an under-parameterized deep decoder lies in a union of low-dimensional subspaces. Hence, taking measurements with sufficiently many i.i.d. Gaussian measurements guarantees that only one such signal is consistent with the measurements.

11.3 Signal reconstruction with an over-parameterized un-trained network

A sufficiently over-parameterized convolutional neural network can fit any single image perfectly, including noise. Thus, at first sight, it may seem surprising that over-parameterized un-trained networks enable accurate signal reconstruction.

Reconstruction is possible since a convolutional network optimized with gradient descent fits a natural image faster than noise. This is illustrated in Figure 11.3, where we show the results of the following experiment:

First we applied gradient descent to fit a clean image (Figure 11.3(b)) and pure noise (Figure 11.3(c)), by minimizing the least-squares loss (11.1) with $\mathbf{A} = \mathbf{I}$, i.e.,

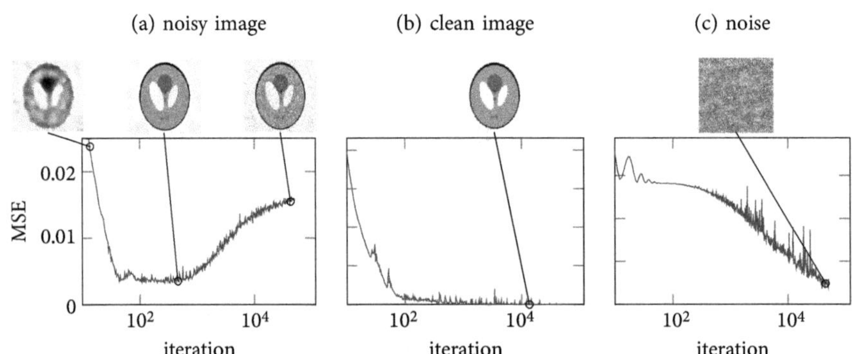

Figure 11.3 Fitting an over-parameterized deep decoder network to (a) a noisy image, (b) a clean image, and (c) pure noise. Here, MSE denotes Mean-Squared Error of the network output with respect to the clean image in (a), and with respect to the clean image in (b), and the noise in (c). While the network can fit the noise due to over-parameterization, it fits natural images with significantly fewer iterations than noise. Hence, when fitting a noisy image, the image component is fitted faster than the noise component, which enables denoising via early stopping. The curves when using other common convolution networks (e.g., a convolutional generator network of a U-net) are qualitatively the same.

$$\mathcal{L}(\boldsymbol{\theta}) = \|G(\boldsymbol{\theta}) - \mathbf{y}\|_2^2 .$$

We plot the least-squares objective as a function of the gradient descent iterations, i.e., we plot $\mathcal{L}(\boldsymbol{\theta}_t)$, where the gradient descent iterations are

$$\boldsymbol{\theta}_{t+1} = \boldsymbol{\theta}_t - \eta \nabla \mathcal{L}(\boldsymbol{\theta}_t).$$

Here, the network $G\colon \mathbb{R}^p \to \mathbb{R}^n$ is the deep decoder architecture (see Figure 11.1), with the number of channels, k, chosen sufficiently large so that the network is over-parameterized (i.e., $p > n$). For fitting the clean image, $\mathbf{y} = \mathbf{x}$ is a clean image, and for fitting the noise, $\mathbf{y} = \mathbf{e}$ is Gaussian noise. For this example, after about 300 iterations the network fits the clean image. At the same time, fitting the Gaussian noise requires around 3000 iterations.

If we apply gradient descent to the least-squares loss

$$\mathcal{L}(\boldsymbol{\theta}) = \|G(\boldsymbol{\theta}) - \mathbf{y}\|_2^2 ,$$

where $\mathbf{y} = \mathbf{x} + \mathbf{e}$ is a noisy image, then the network first fits the image part of the noisy image and only later the noise part. This is shown in Figure 11.3(a)), where we plot the MSE of representing the signal, i.e., $\|G(\boldsymbol{\theta}_t) - \mathbf{x}\|_2^2$ as a function of the gradient descent iterations. Thus, early stopping the gradient descent iterations at about $t = 300$ iterations denoises the image relatively well, since after about $t = 300$ iterations the signal has been fitted well but little of the noise has been fitted.

For denoising with an over-parameterized un-trained network, regularization via early stopping is critical for performance, since with enough iterations the network fits the entire noisy image. The early stopping time plays an analogous role to the number of parameters for image reconstruction with an under-parameterized neural network: With more iterations, gradient descent fits the image part better, but also fits more of the noise.

Empirically, un-trained neural networks often perform best in the over-parameterized regime. Several variants of convolutional generator networks work well, including the deep decoder, a U-net, and a convolutional generator network [HH19, DH21, UVL18].

11.3.1 Provable denoising with over-parameterized convolutional networks

In order to understand the empirical observation that a convolutional network fits a natural image faster than noise, and thus early stopping can denoise natural images, we examine this observation theoretically in this more technical section.

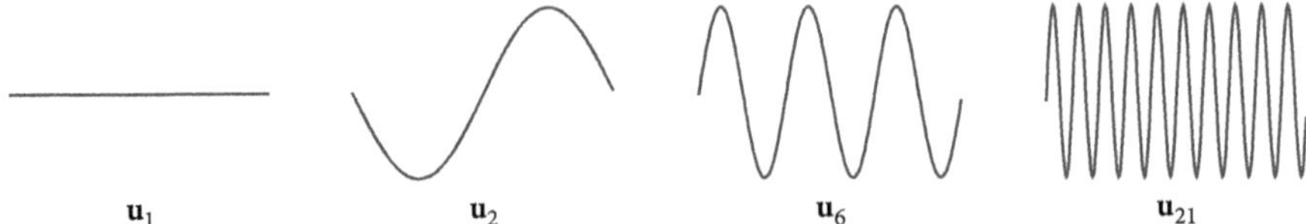

Figure 11.4 The 1st, 2nd, 6th, and 21st trigonometric basis functions in dimension $n = 300$.

As a model for natural images, we consider smooth signals. We say a signal $\mathbf{x} \in \mathbb{R}^n$ is *q-smooth* if it can be represented as a linear combination of the q first trigonometric basis functions (illustrated in Figure 11.4). Natural images are commonly modelled as smooth signals since the power spectrum (i.e., the energy distribution by frequency) of a natural image decays rapidly from low frequencies to high frequencies (see for example [SO01, Fig. 4]).

We consider a two-layer convolutional network of the form

$$G(\boldsymbol{\theta}) = \text{relu}(\mathbf{U}_0\mathbf{B}_0\boldsymbol{\Theta}_0)\boldsymbol{\theta}_1, \tag{11.3}$$

where $\mathbf{U}_0 \in \mathbb{R}^{n\times\frac{n}{2}}$ is a linear operator implementing a convolution with a fixed upsampling operator, and $\boldsymbol{\Theta}_0 \in \mathbb{R}^{k_0\times k_0}$ is a parameter matrix forming linear combinations of the channels $\mathbf{U}_0\mathbf{B}_0$. Finally, we apply a ReLU non-linearity and again form linear combinations through multiplication with the parameter vector $\boldsymbol{\theta}_1 \in \mathbb{R}^{k_0}$, which yields the output image. The parameters of the network are the weights $\boldsymbol{\theta} = (\boldsymbol{\Theta}_0, \boldsymbol{\theta}_1)$. This is a two-layer deep decoder network.

The result stated below relies on the insight that the behaviour of large over-parameterized neural networks is dictated by the spectral properties of its Jacobian mapping at initialization. The Jacobian of the function $G: \mathbb{R}^p \to \mathbb{R}^n$ is the matrix $\mathbf{J} \in \mathbb{R}^{n\times p}$, whose (i, j)-th entry is equal to the derivative of the i-th output value with respect to the j-th input value. The left-singular values of the expected Jacobian of this convolutional network at initialization are the trigonometric basis functions $\mathbf{u}_1, \ldots, \mathbf{u}_n$ (with expectation taken over the random initialization). Provided that the fixed convolutional filter of the convolution operation in the deep decoder network is relatively narrow (which it is in practice), the associated singular values $\sigma_1 > \sigma_2 > \ldots > \sigma_n$ decay rapidly, so that large singular values are associated with low-frequency trigonometric basis functions and small singular values are associated with high-frequency basis functions.

The following result shows that for un-trained convolutional networks, gradient descent fits the components of the noisy measurement $\mathbf{y} = \mathbf{x} + \mathbf{e}$ that align with the trigonometric basis functions at speeds determined by the associated singular values.

Theorem 10 (Denoising guarantees with over-parameterized networks [HS20b, Thm. 2]) *Assume that* $\mathbf{x}$ *is a q-smooth signal, and let* $\mathbf{e}$ *be an arbitrary noise vector. Suppose that we fit the randomly initialized two-layer convolutional network (11.3) via gradient descent with step size* $\eta \leq \frac{1}{\sigma_1^2}$ *for t iterations to minimize the least-squares loss* $\mathcal{L}(\boldsymbol{\theta}) = \|G(\boldsymbol{\theta}) - \mathbf{y}\|_2^2$ *with* $\mathbf{y} = \mathbf{x} + \mathbf{e}$. *Suppose that the network is sufficiently wide, specifically,* $k \geq \Omega(\frac{n}{\epsilon^4})$, *for some* $\epsilon > 0$. *Then the estimate of the un-trained network based on the t-th iterate* $\boldsymbol{\theta}_t$ *obeys, with high probability over the random initialization,*

$$\|G(\boldsymbol{\theta}_t) - \mathbf{x}\|_2 \leq (1 - \eta\sigma_q^2)^t \|\mathbf{x}\|_2 + \left(\sum_{i=1}^{n}((1 - \eta\sigma_i^2)^t - 1)^2 \mathbf{e}^T \mathbf{u}_i\right)^{1/2} + \epsilon, \tag{11.4}$$

where $\{\sigma_i\}_{i=1}^n$ *are the singular values described above.*

In this result, ϵ is an error term that becomes negligible if the network is sufficiently wide. The first term in the error bound (11.4) is the error for fitting the signal, and the second term corresponds to the noise fitted after t iterations. After sufficiently many iterations, $(1 - \eta\sigma_q^2)^t$ is small, and thus the signal fitting error is small. At the same time, after such a number of iterations, only the components of the noise that align with (roughly) the q-many lowest frequency trigonometric basis functions are fitted, provided that the singular values decay sufficiently fast.

A consequence of this result (see [HS20b, Thm. 1]) is that there is an optimal number of iterations such that for denoising a signal corrupted with Gaussian noise $\mathbf{e} \sim \mathcal{N}(0, \mathbf{I})$, the estimate based on early-stopped gradient descent obeys

$$\|G(\boldsymbol{\theta}_t) - \mathbf{x}\|_2 \leq O\left(\frac{q}{n}\right).$$

This guarantees that only a fraction $\frac{q}{n}$ of the noise energy is fitted, and much of the rest of the noise that lies outside of the signal subspace spanned by the q-lowest frequency trigonometric basis functions is filtered out. That is, up to a constant factor, optimal for denoising a q-smooth signal with Gaussian noise.

Here we discussed denoising, but an un-trained convolutional neural network can also provably estimate a smooth signal well from few random measurements [HS20a].

11.3.2 Appealing to linear least-squares

Next, we develop intuition for Theorem 10 by studying a linear least-squares problem. Minimizing the loss in Equation (11.1) is a non-linear least-squares problem, since the network $G(\boldsymbol{\theta})$ is a non-linear function of its parameters. In order to understand why fitting a convolutional generator with gradient descent can fit some components of the signal faster than others, we study the dynamics of gradient descent applied to a linear least-squares problem. We demonstrate how early stopping can lead to denoising capabilities even with a simple linear model.

We consider a least-squares problem of the form

$$\mathcal{L}(\boldsymbol{\theta}) = \frac{1}{2} \|\mathbf{J}\boldsymbol{\theta} - \mathbf{y}\|_2^2,$$

and study the estimate $\mathbf{J}\boldsymbol{\theta}_k$ obtained by running gradient descent with a constant step size η starting at $\boldsymbol{\theta}_0 = 0$ for k iterations. The iterates of gradient descent are

$$\boldsymbol{\theta}_{t+1} = \boldsymbol{\theta}_t - \eta \nabla \mathcal{L}(\boldsymbol{\theta}_t), \quad \nabla \mathcal{L}(\boldsymbol{\theta}) = \mathbf{J}^T(\mathbf{J}\boldsymbol{\theta} - \mathbf{y}).$$

The following simple proposition, proven at the end of this subsection, characterizes the trajectory of gradient descent.

Proposition 12 *Let* $\mathbf{J} \in \mathbb{R}^{n \times p}$ *be a matrix with left-singular vectors* $\mathbf{u}_1, \ldots, \mathbf{u}_n \in \mathbb{R}^n$ *and corresponding singular values* $\sigma_1 \geq \sigma_2 \geq \ldots \geq \sigma_n$. *Then the residual after* t *steps,* $\mathbf{r}_t = \mathbf{y} - \mathbf{J}\boldsymbol{\theta}_t$, *of gradient descent starting at* $\boldsymbol{\theta}_0 = 0$ *is*

$$\mathbf{r}_t = \sum_{i=1}^{n} \mathbf{u}_i \mathbf{u}_i^T \mathbf{y} (1 - \eta \sigma_i^2)^t.$$

Suppose that the signal $\mathbf{y}$ lies in the column span of $\mathbf{J}$, and that the step size is chosen sufficiently small (i.e., $\eta \leq 1/\|\mathbf{J}\|^2$). Then, by Proposition 12, gradient descent converges to a zero-loss solution and thus fits the signal perfectly. More importantly for our discussion, gradient descent fits the components of $\mathbf{y}$ corresponding to large singular values faster than it fits the components corresponding to small singular values.

To explicitly show how this observation enables regularization via early-stopped gradient descent, suppose our goal is to find a good estimate of a signal $\mathbf{x}$ from a noisy observation

$$\mathbf{y} = \mathbf{x} + \mathbf{e},$$

where the signal $\mathbf{x}$ lies in a signal subspace that is spanned by the q leading left-singular vectors of $\mathbf{J}$. Then, by Proposition 12, the signal estimate after t iterations, $\mathbf{J}\boldsymbol{\theta}_t$, obeys

$$\|\mathbf{J}\boldsymbol{\theta}_t - \mathbf{x}\|_2 \leq (1 - \eta\sigma_q^2)^t \|\mathbf{x}\|_2 + E(\mathbf{e}), \quad E(\mathbf{e}) := \sqrt{\sum_{i=1}^{n}((1 - \eta\sigma_i^2)^t - 1)^2 \langle \mathbf{u}_i, \mathbf{e}\rangle^2}. \tag{11.5}$$

Thus, after a few iterations most of the signal has been fitted (i.e., $(1 - \eta\sigma_q)^t$ is small). Furthermore, if we assume that the ratio σ_q/σ_{q+1} is sufficiently large so that the spread between the two singular values separating the signal subspace from the rest is sufficiently large, most of the noise outside the signal subspace has not been fitted (i.e., $((1 - \eta\sigma_i^2)^t - 1)^2 \approx 0$ for $i = q + 1, ..., n$).

In particular, suppose the noise vector has a Gaussian distribution given by $\mathbf{e} \sim \mathcal{N}(0, \frac{\varsigma^2}{n}\mathbf{I})$. Then $E(\mathbf{e}) \approx \varsigma\sqrt{\frac{q}{n}}$ so that after order $t = \log(\epsilon)/\log(1 - \eta\sigma_q^2)$ iterations, with high probability,

$$\|\mathbf{J}\boldsymbol{\theta}_t - \mathbf{x}\|_2 \leq \epsilon \|\mathbf{x}\|_2 + c\varsigma\sqrt{\frac{q}{n}}.$$

This demonstrates, that provided the signal lies in a subspace spanned by the leading singular vectors, early-stopped gradient descent reaches the optimal denoising rate of $\varsigma\sqrt{q/n}$ after a few iterations. See Figure 11.5 for a numerical example demonstrating this phenomena.

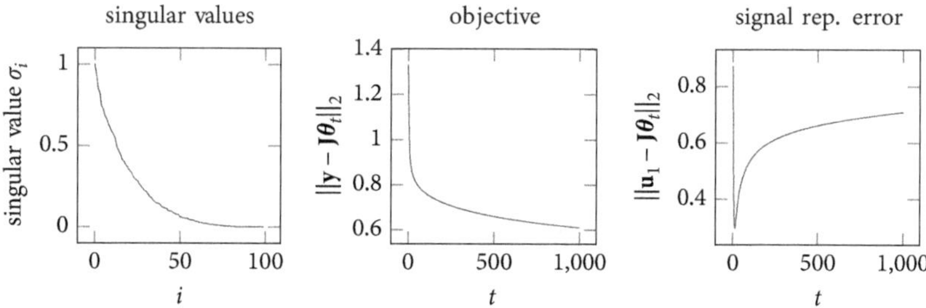

Figure 11.5 Gradient descent for minimizing the least-squares problem of minimizing the objective $\|\mathbf{y} - \mathbf{J}\boldsymbol{\theta}_t\|_2^2$, where the Jacobian $\mathbf{J} \in \mathbb{R}^{100\times100}$ has decaying singular values, and the measurement $\mathbf{y}$ is the sum of a signal component, which is the leading singular vector $\mathbf{u}_1$ of the Jacobian $\mathbf{J}$, and a noisy component $\mathbf{e} \sim \mathcal{N}(0, (1/n)\mathbf{I})$, i.e., $\mathbf{y} = \mathbf{u}_1 + \mathbf{e}$. **Left panel:** The decaying singular values. **Middle panel:** least-square objective. **Right panel:** The signal component $\mathbf{u}_1$ is fitted significantly faster than the other components (right panel), but then overfits to the noise, thus early stopping enables denoising.

Proof of Proposition 12: The residual of gradient descent at iteration t is

$$\begin{aligned}\mathbf{r}_t &= \mathbf{y} - \mathbf{J}\boldsymbol{\theta}_t \\ &= \mathbf{y} - \mathbf{J}(\boldsymbol{\theta}_{t-1} - \eta\mathbf{J}^T(\mathbf{J}\boldsymbol{\theta}_{t-1} - \mathbf{y})) \\ &= (\mathbf{I} - \eta\mathbf{J}\mathbf{J}^T)(\mathbf{y} - \mathbf{J}\boldsymbol{\theta}_{t-1}) \\ &= (\mathbf{I} - \eta\mathbf{J}\mathbf{J}^T)^t(\mathbf{y} - \mathbf{J}\boldsymbol{\theta}_0) \\ &= (\mathbf{I} - \eta\mathbf{J}\mathbf{J}^T)^t\mathbf{y} \\ &= (\mathbf{I} - \eta\mathbf{U}\boldsymbol{\Sigma}^2\mathbf{U}^T)^t\mathbf{y},\end{aligned}$$

where we used that $\boldsymbol{\theta}_0 = 0$ and the SVD $\mathbf{J} = \mathbf{U}\boldsymbol{\Sigma}\mathbf{V}^T$. Expanding $\mathbf{y}$ in terms of the singular vectors $\mathbf{u}_i$ (i.e., the columns of $\mathbf{U}$), as $\mathbf{y} = \sum_i \mathbf{u}_i \langle \mathbf{u}_i, \mathbf{y}\rangle$, and noting that $(\mathbf{I} - \eta\mathbf{U}\boldsymbol{\Sigma}^2\mathbf{U}^T)^t = \sum_i(1 - \eta\sigma_i^2)^t\mathbf{u}_i\mathbf{u}_i^T$, we get

$$\mathbf{r}_t = \sum_i (1 - \eta\sigma_i^2)^t\mathbf{u}_i \langle \mathbf{u}_i, \mathbf{y}\rangle,$$

as desired.

Proof of Equation (11.5) *By Proposition 12 and using that* $\mathbf{x}$ *lies in the signal subspace*

$$\mathbf{x} - \mathbf{J}\boldsymbol{\theta}_t = \sum_{i=1}^{q} \mathbf{u}_i(1 - \eta\sigma_i^2)^t \langle \mathbf{u}_i, \mathbf{x}\rangle + \sum_{i=1}^{n} \mathbf{u}_i((1 - \eta\sigma_i^2)^t - 1) \langle \mathbf{u}_i, \mathbf{e}\rangle.$$

By the triangle inequality,

$$\begin{aligned}\|\mathbf{x} - \mathbf{J}\boldsymbol{\theta}_t\|_2 &\le \left\|\sum_{i=1}^{q} \mathbf{u}_i(1 - \eta\sigma_i^2)^t \langle \mathbf{u}_i, \mathbf{x}\rangle\right\|_2 + \left\|\sum_{i=1}^{n} \mathbf{u}_i((1 - \eta\sigma_i^2)^t - 1) \langle \mathbf{u}_i, \mathbf{e}\rangle\right\|_2 \\ &\le (1 - \eta\sigma_q^2)^t \|\mathbf{x}\|_2 + \sqrt{\sum_{i=1}^{n}((1 - \eta\sigma_i^2)^t - 1)^2 \langle \mathbf{u}_i, \mathbf{e}\rangle^2},\end{aligned}$$

where the second inequality follows by using orthogonality of $\mathbf{u}_i$ *and by using* $(1 - \eta\sigma_i^2) \le 1$*, from* $\eta \le 1/\sigma_{\max}^2$*. This concludes the proof of Equation (11.5).*

11.4 Chapter notes

In this chapter we have discussed un-trained neural networks for image reconstruction. Un-trained neural networks can perform quite well, as illustrated in Figure 11.6: They typically yield better results than sparsity-based reconstruction,

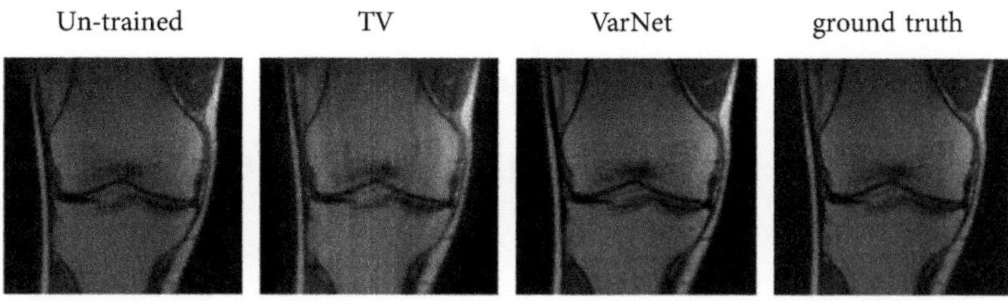

Figure 11.6 Sample reconstructions for an un-trained network optimized for MRI, total-variation regularized least-squares and the end-to-end variational network (VarNet) for a validation image from multi-coil knee measurements (4x accelerated) from the fastMRI dataset. It can be seen that the un-trained network performs significantly better than the sparse reconstruction-based approach, but worse than the network trained in a supervised fashion (VarNet). This example is illustrative of the relative behaviour of those three algorithms for accelerated MRI.

which is also an un-trained reconstruction approach, but perform worse than supervised approaches, if sufficient training data is available for the supervised approach. It is not surprising that a supervised approach gives better reconstructions, since it can learn an effective image prior from data. It is quite remarkable, however, that the un-trained network performs so well relative to total-variation regularized least-squares.

The fact that un-trained networks perform well for image reconstruction also illustrates that the convolutional networks used for image reconstruction are very good image priors by themselves, even without any training.

In terms of computational reconstruction cost, un-trained networks are similarly expensive as regularized least-squares, and significantly more expensive than networks trained in a supervised fashion, such as the variational network.

This makes un-trained neural network-based approaches an interesting tool for image reconstruction in a setup where little training data is available to use a supervised approach or a reconstruction approach based on generative prior.

12
Coordinate-based multi-layer perceptrons

Coordinate-based neural representations for images, 3D shapes, and other signals have recently emerged as an alternative to traditional discrete representations, such as sparse representations and convolutional neural networks. They are also known as implicit neural representations and were originally introduced for representing scenes and for performing view synthesis [Mil+20] and surface reconstruction [Wil+19] for computer vision and graphics applications. Coordinate-based representations enable to represent a signal efficiently in high dimensions. They can also be useful for image representation and reconstruction [Tan+20; Dup+21], and are particularly useful for representing and working with high-dimensional signals, such as high-resolution images and volumes.

Coordinate-based networks represent a colour image by mapping a pixel-coordinate (x, y) to a corresponding RGB intensity value. More generally, coordinate-based networks can be used to represent signals by mapping a coordinate to a signal value (e.g., an (x, y, z)-coordinate to a voxel value, or a spacial-temporal coordinate (x, y, t) to a pixel value for an image of an object that changes across time).

Important for the performance of coordinate-based networks for imaging tasks is the network architecture. As we will see below, a simple ReLU-multi-layer-perceptron (MLP) directly operating on the input coordinates is not efficient for representing images and 3D objects, but a ReLU-MLP operating on coordinates transformed with a sinusoidal "positional encoding" or with random Fourier-features performs very well [Mil+20; Tan+20].

In this chapter, we discuss coordinate-based neural networks for solving image reconstruction problems. We discuss a particular architecture, a multi-layer perceptron with Fourier-feature inputs, which works well for representing images and 3D-objects. This architecture can be used as an image prior in an analogous way as un-trained networks discussed in Chapter 11. As of now, this architecture allows to solve image reconstruction problems (e.g., denoising or compressive sensing) reasonably well, but is often not competitive with the un-trained convolutional neural networks discussed in Chapter 11, at least for traditional image reconstruction tasks like denoising and accelerated MRI. However, the method is conceptually interesting and gives a new perspective on how signals can be represented and manipulated with neural networks.

Deep Learning for Computational Imaging. Reinhard Heckel, Oxford University Press. © Reinhard Heckel (2025).
DOI: 10.1093/oso/9780198947172.003.0012

12.1 Coordinate-based neural network architectures

We consider coordinate-based neural networks, i.e., neural networks that map a coordinate $\mathbf{c} \in \mathbb{R}^d$ (with $d = 2$ for images and $d = 3$ for a volume) to a real number representing a greyscale value, or to three real numbers representing an RGB value. See Figure 12.1 (left panel) for an illustration. As discussed next, a fully-connected network with the input coordinates transformed by a Fourier-feature map is a good image model. A Fourier-feature map with parameters $\mathbf{W} = [\mathbf{w}_1, \ldots, \mathbf{w}_k]^T \in \mathbb{R}^{k \times d}$ is defined as

$$\gamma_{\mathbf{W}}(\mathbf{c}) = \frac{1}{\sqrt{k}} \left[\gamma_1(\mathbf{c}), \ldots, \gamma_k(\mathbf{c})\right], \quad \gamma_i(\mathbf{c}) = [\cos(\mathbf{w}_i^T \mathbf{c}), \sin(\mathbf{w}_i^T \mathbf{c})]. \tag{12.1}$$

Here, k determines the dimension of the Fourier-feature map (i.e., $\gamma_i(\mathbf{c}) \in \mathbb{R}^{2k}$). When used as a feature map, the weights $\mathbf{W}$ are randomly initialized as i.i.d. Gaussians and are fixed (i.e., they are not trainable parameters).

To illustrate that a fully-connected network with the input coordinates transformed by a Fourier-feature map is a good image model, we discuss the following four network architectures:

- ReLU-MLP: A standard multi-layer-perceptron (MLP) with ReLU-activation functions:

$$f(\mathbf{c}) = \mathbf{W}_\ell^T \pi_{\ell-1} \circ \pi_{\ell-2} \circ \ldots \circ \pi_0(\mathbf{c}), \tag{12.2}$$

  where $\pi_i(\mathbf{c}) = \sqrt{\frac{2}{k}}\xi(\mathbf{W}_i \mathbf{c} + \mathbf{b}_i)$ is a linear layer parameterized by the weight matrix $\mathbf{W}_i$, bias $\mathbf{b}_i$, and ReLU-activation function ξ. Here, $f \circ g$ denotes function decomposition, i.e., applying function g first and then applying function f to the result.
- FF+MLP: A ReLU-MLP that, instead of taking the original coordinates $\mathbf{c}$ as input, takes a random Fourier-feature map of the coordinates as input. This architecture, proposed and studied by Tancik et al. [Tan+20], is very similar to the Neural Radiance Fields (NeRF) architecture by Mildenhall et al. [Mil+20], which uses a Fourier-feature map separately for each coordinate of $\mathbf{c}$, called positional encoding.
- FF: A Fourier-feature MLP which concatenates Fourier-feature maps $f(\mathbf{c}) = \mathbf{W}_\ell^T \gamma_{\mathbf{W}_{\ell-1}} \circ \ldots \circ \gamma_{\mathbf{W}_0}(\mathbf{c})$, where the parameters $\mathbf{W}_\ell, \ldots, \mathbf{W}_1$ are trainable and $\mathbf{W}_0$ is fixed. This is a version of the SIREN network, which is an MLP with sine activation functions proposed by Sitzmann et al. [Sit+20]. The SIREN network and the FF are almost equivalent, they only differ in the parameterization of the Fourier-features.

- FF+NormReLU: Same as FF+MLP, but we normalize the outputs of each MLP π_i with $\frac{\pi_i - \text{mean}(\pi_i)}{\text{std}(\pi_i)}$. This is similar to popular normalization layers, and aids the optimization. This network will turn out to perform similar to FF and FF+NormReLU.

12.2 Coordinate-based image representations

Throughout this book, we have seen that good image or signal models enable to represent an image or signal with few parameters. In order to understand which architecture is a good image model, we therefore start by representing an image with an under-parameterized network, i.e., a network that has fewer parameters than the image has pixels, and measure the representation error.

Let $\mathbf{x} \in \mathbb{R}^{W \times H}$ be the pixel values of an image of width W and height H, and let $f_{\theta}: \mathbb{R}^2 \to \mathbb{R}$ be one of the four coordinate-based image representations introduced in the previous section, with parameters $\boldsymbol{\theta}$. We fit an image $\mathbf{x}$ by applying gradient descent to the loss

$$\mathcal{L}(\boldsymbol{\theta}) = \frac{1}{2} \sum_{w,h} (x_{w,h} - f_{\theta}(\mathbf{c}_{w,h}))^2,$$

where the sum is over all pixels of the image indexed by width and height (w, h) and the coordinate vectors $\mathbf{c}_{w,h} \in \mathbb{R}^2$ lie on a rectangular grid over $[-1, 1]^2$ (i.e., $\mathbf{c}_{w,h} \in \{(-1, -1), (-1 + 2/w, -1), \ldots, (-1 + 2(w-1)/w, -1 + 2(h-1)/h\})$.

Figure 12.1 (right panel) shows how well the different architectures, each with three hidden layers and an identical number of parameters, perform at fitting a single image.

It can be seen that ReLU-MLP is significantly outperformed (by almost 10 dB) by FF+MLP, which in turn is significantly outperformed (again by almost 10 dB) by FF and FF+NormReLU. Those comparative results are representative for natural images.

This experiment demonstrates that multi-layer perceptrons with Fourier-feature inputs are very good at representing images.

This experiment also demonstrates that normalization is critical for performance. To see this, recall that our experiment shows that FF+MLP is significantly outperformed by FF and FF+NormReLU; this only happens if the network is moderately deep. For a network with a single hidden layer, all networks with Fourier-feature inputs perform equally well. The reason why FF+NormReLU and FF outperform FF+MLP can be attributed to the explicit (FF+NormReLU) and implicit (FF) normalization, which aids the optimization of the networks. To see the implicit normalization property, note that $\|\gamma_{\mathbf{W}}(\mathbf{c})\|_2 = 1$, regardless of the input $\mathbf{c}$ and the weights $\mathbf{W}$. Thus, the Fourier-feature maps in deeper layers have a

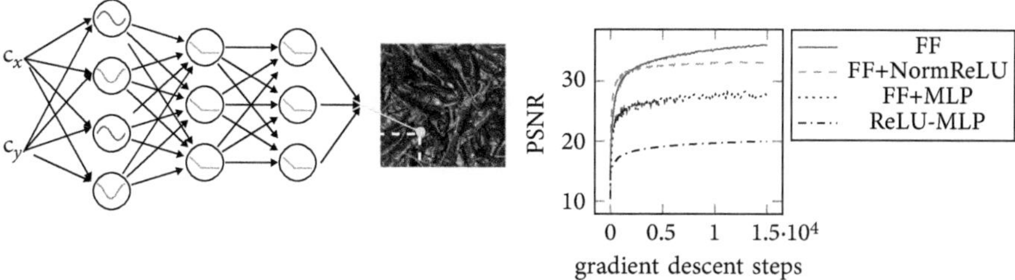

Figure 12.1 **Left**: A coordinate-based network encodes an image pixel-by-pixel. The input to the network is a coordinate vector $\mathbf{x} = [c_x, c_y]$ and the output the pixel value. **Right**: Training curves for fitting a 512×512 greyscale image with different coordinate-based neural networks, each with half the number of parameters than pixels and with three hidden layers. Fourier-features as inputs is critical for performance, and using layers that normalize (through explicit normalization or through Fourier-features within the network) is also critical for performance.

similar effect as explicit normalization layers that are known to be important for optimizing deep networks.

12.3 MLPs with Fourier-feature inputs fit low- before high-frequency components

We next show that coordinate-based MLPs with Fourier-feature inputs fit low-frequency components of a signal before high-frequency components, analogous to convolutional neural networks. Thus, coordinate-based MLPs with Fourier-feature inputs have a preference towards low-frequency signals. This sheds light on why these networks are good image priors: Natural images are well-approximated by smooth signals, as discussed before (see for example Simoncelli and Olshausen (2001) [SO01]).

Our discussion will also reveal that the variance of the random features primarily determines the frequency decay: A smaller variance leads to a higher preference towards smooth signals. The weights of the random Fourier-features $\mathbf{W}_0$ are initialized as i.i.d. $\mathcal{N}(0, \sigma_0^2)$ random variables, and thus the variance is set by σ_0^2.

12.3.1 Background on linear approximation of neural networks

To understand network architectures theoretically, we study linear approximations of an MLP with Fourier-feature inputs. We start by providing background on this technique. Theorem 10 in Chapter 11 for un-trained convolutional networks is also based on such a linear approximation.

Let $f_{\boldsymbol{\theta}}: \mathbb{R}^d \to \mathbb{R}$ be a coordinate-based network with parameter vector $\boldsymbol{\theta} \in \mathbb{R}^p$ mapping a d-dimensional pixel-coordinate to a single greyscale output. Let $\mathbf{f}(\boldsymbol{\theta}) \in \mathbb{R}^n$ be the predictions of the network for the n pixels on an equispaced (rectangular) grid. The neural-tangent-kernel literature [JGH18, Du+18, OS20] established that if the network is initialized randomly with a sufficiently small initialization $\boldsymbol{\theta}_0$, and if it is sufficiently wide, then the outputs of the network are well-approximated by a linear function around its initialization $\boldsymbol{\theta}_0$, i.e.,

$$\begin{aligned}\mathbf{f}(\boldsymbol{\theta}) &\approx \mathbf{f}(\boldsymbol{\theta}_0) + \mathbf{J}(\boldsymbol{\theta} - \boldsymbol{\theta}_0) \\ &\approx \mathbf{J}\boldsymbol{\theta}.\end{aligned}$$

Here, $\mathbf{J} \in \mathbb{R}^{n \times p}$ is the Jacobian of the network at initialization, and the second approximation holds provided that the initialization is sufficiently small, so that $\mathbf{f}(\boldsymbol{\theta}_0) \approx 0$ and $\mathbf{J}\boldsymbol{\theta}_0 \approx 0$. Let $\mathbf{J} = \mathbf{U}\boldsymbol{\Sigma}\mathbf{V}^T$ be the singular value decomposition of the Jacobian. When fitting the data of a sufficiently wide network with gradient descent, the iterates remain in the area around the initialization in which the outputs of the network are well-approximated with gradient descent. With the approximation above, the predictions after t iterations of gradient descent applied to fitting the network to an image $\mathbf{x}$ by minimizing the least-squares loss $\mathcal{L}(\boldsymbol{\theta}) = \frac{1}{2} \|\mathbf{f}(\boldsymbol{\theta}) - \mathbf{x}\|_2^2$ are given by

$$\mathbf{f}(\boldsymbol{\theta}_t) \approx \sum_{i=1}^{n} \mathbf{u}_i(\mathbf{u}_i^T \mathbf{x})(1 - (1 - \eta\sigma_i^2)^t). \tag{12.3}$$

This relation shows that the predictions of the network are determined by the left-singular vectors of the Jacobian and the corresponding singular values. If the signal $\mathbf{x}$ lies in the span of the left-singular vectors, then at convergence of gradient descent, the network represents the signal perfectly (i.e., $\mathbf{x} = \mathbf{f}(\boldsymbol{\theta}_\infty)$). Moreover, components of the signal $\mathbf{x}$ associated with large singular values are fitted faster than components associated with small singular values.

As the network becomes wide, those singular vectors are approximately equal to the singular vectors of an associated kernel defined as

$$\mathbf{U}\boldsymbol{\Sigma}^2\mathbf{U}^T \approx \mathbf{K} = \mathbb{E}_{\boldsymbol{\theta}_0}\left[\mathbf{J}(\boldsymbol{\theta}_0)\mathbf{J}^T(\boldsymbol{\theta}_0)\right], \tag{12.4}$$

where expectation is over the random initialization of the network. This expectation and the associated singular vectors and values can often explicitly be computed.

While in practice, the network is not infinitely wide, and thus the approximation above is not exact, the linear approximation is useful to understand which priors a given network implements. For example, as Theorem 10 in Chapter 11 shows, untrained convolutional networks have a preference towards fitting low-frequency components of a signal. Moreover, Tancik et al. [Tan+20] have shown, by analysing an infinitely wide coordinate-based network with Fourier-feature inputs that "passing input points through a simple Fourier feature mapping enables a multilayer perceptron (MLP) to learn high-frequency functions in low-dimensional problem domains", thus effectively showing that Fourier-feature maps also induce a bias towards fitting low-frequency components.

12.3.2 Gradient descent dynamics of MLPs with Fourier-feature inputs

As shown in the previous subsection, for wide networks, the dynamics of fitting an image by minimizing a least-squares loss with gradient descent are determined by the singular vector and associated singular values of the kernel (12.4). The following statement characterizes the singular vectors and associated singular values of an MLP with random Fourier-feature inputs.

Theorem 11 *Consider a multi-layer MLP with Fourier-feature input $\gamma_{\mathbf{W}}(\mathbf{x})$ with the weights $\mathbf{W}$ drawn i.i.d. from a zero-mean Gaussian with variance σ_0^2. The associated kernel (12.4) is given by*

$$\mathbf{K}_{ij} = h\left(\exp\left(-\frac{1}{2}\sigma_0^2 \left\|\mathbf{c}_i - \mathbf{c}_j\right\|_2^2\right)\right), \tag{12.5}$$

where h is a kernel function depending on the MLP architecture and initialization.

Theorem 11 implies the following bias towards fitting smooth signals. The kernel matrix $\mathbf{K}$ is banded Toeplitz, and thus is close to a circulant matrix. For equispaced inputs to the network, i.e., $c_i = 1/n, \ldots, n/n$, the singular vectors of the kernel matrix are plotted in Figure 12.2 and can be seen to be close to the trigonometric basis vectors, which are the singular vectors of a circulant matrix. The associated singular values for the ReLU-MLP architecture are plotted in Figure 12.2, left panel. The singular values associated with low-frequency basis functions are large, and the others are small. This, in turn, implies that when fitting data with a multi-layer MLP with Fourier-features, the low-frequency components of a signal are fitted faster than the high-frequency components.

Thus, fitting a network to an image and early stopping the iterations induces a preference towards low-frequency signals. This provides an explanation why coordinate-based networks with Fourier-features are good image models, since natural images have most of the energy on low frequencies [SO01].

Tancik et al. [Tan+20] established a related result, specifically that using Fourier-features of the form

$$\gamma(\mathbf{x}) = [a_1 \cos(\mathbf{w}_1^T \mathbf{x}), a_1 \sin(\mathbf{w}_1^T \mathbf{x}), \ldots, a_m \cos(\mathbf{w}_m^T \mathbf{x}), a_m \sin(\mathbf{w}_m^T \mathbf{x})]$$

as an input yields a shift-invariant kernel, i.e., $[\mathbf{K}]_{ij} = h(\mathbf{c}_i - \mathbf{c}_j)$, where h is a kernel depending on the MLP architecture and initialization. Tancik et al. [Tan+20] then showed that by tuning the coefficients a_i and setting $\mathbf{b}_j = j$, one can tune the properties of the kernel, in particular their frequency decay.

12.3.3 Image resolution and variance parameter of the Fourier-features

The variance parameter σ_0^2 of the Fourier-feature input determines how much the image values of nearby coordinates are related, and thus effectively determines the image resolution. At first sight, coordinate-based neural networks represent images at any resolution. However, if σ_0^2 is small relative to the distance between two coordinates $\mathbf{c}_i$ and $\mathbf{c}_j$, specifically if the term $\frac{1}{2}\sigma_0^2 \left\| \mathbf{c}_i - \mathbf{c}_j \right\|_2^2$ in Equation (12.5) is large, then the pixel values corresponding to the two coordinates $\mathbf{c}_i$ and $\mathbf{c}_j$ have very similar values. Thus, the variance parameters determine at what scale pixel values are interpolated and thus effectively determine the resolution at which an MLP with Fourier-feature inputs can represent a signal.

To see this, we plot in Figure 12.3 a spoke target with 512×512 pixels, that we fitted with a Fourier-feature MLP with different values for the variance parameter σ_0^2. We choose the variance parameter so that the factor in Equation (12.5) between two neighbouring pixels ($\mathbf{c}_i$ and $\mathbf{c}_j$ are two neighbouring pixels) takes the values

$$\frac{1}{2}\sigma_0^2 \left\| \mathbf{c}_i - \mathbf{c}_j \right\|_2^2 = \frac{1}{2}\sigma_0^2 (2/512)^2 = 1.2\text{e-}4, 2.7\text{e-}4, 4.8\text{e-}4, 2\text{e-}3.$$

It can be seen that if the variance parameter is too small, neighbouring pixels are interpolated too much and the image becomes very smooth, while for larger variance parameters, the pixels of the image are well-resolved.

12.4 Coordinate-based neural representations for image reconstruction problems

Coordinate-based neural representation can be used analogously as un-trained neural networks discussed in Chapter 11 for image reconstruction.

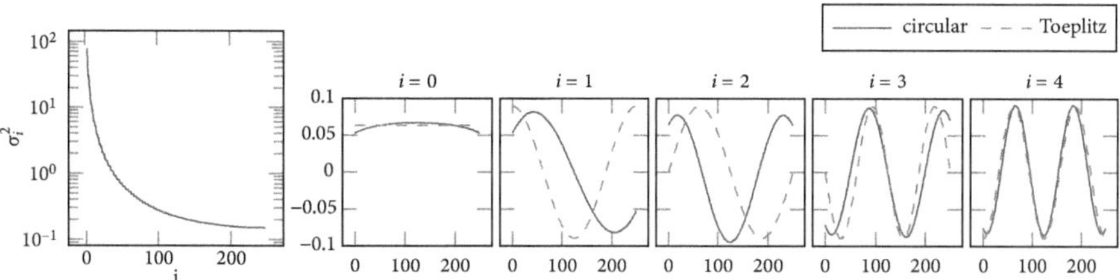

Figure 12.2 Left: Singular values of the Jacobian induced by a network with Fourier-feature inputs followed by a ReLU-MLP. **Right**: The associated singular vectors (Toeplitz) and for comparison, the trigonometric basis functions which are closely related.

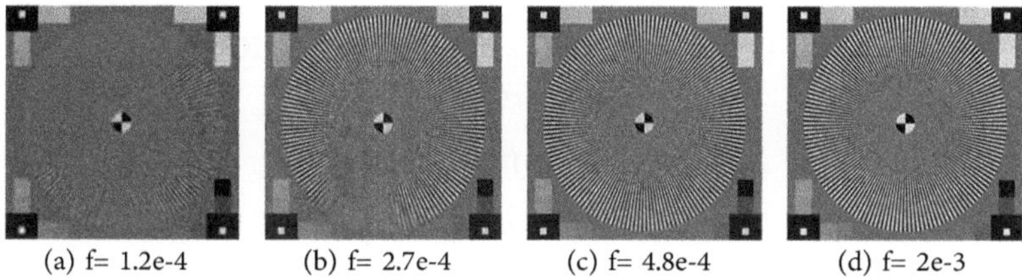

(a) f= 1.2e-4 (b) f= 2.7e-4 (c) f= 4.8e-4 (d) f= 2e-3

Figure 12.3 A spoke target fitted with a Fourier-feature MLP with different values for the variance parameter σ_0^2, so that the factor in Equation (12.5) between two neighbouring pixels $\mathbf{c}_i$ and $\mathbf{c}_j$ takes on the values $\frac{1}{2}\sigma_0^2 \left\|\mathbf{c}_i - \mathbf{c}_j\right\|_2^2 = 1.2\text{e-}4$, 2.7e-4, 4.8e-4, 2e-3. For small values of the variance parameter, neighbouring pixels are interpolated while for large values, pixels are well-resolved.

Specifically, suppose our goal is to reconstruct an image $\mathbf{x}$ from noisy linear measurements $\mathbf{y} = \mathbf{A}\mathbf{x} + \mathbf{e}$. Let $f_{\boldsymbol{\theta}}: \mathbb{R}^2 \to \mathbb{R}$ be a coordinate-based network with parameters $\boldsymbol{\theta}$, and let $\mathbf{f}(\boldsymbol{\theta})$ be the predictions of the network on a rectangular grid corresponding to the pixels of the image. We reconstruct an image by first fitting the image with gradient descent to the loss

$$\mathcal{L}(\boldsymbol{\theta}) = \|\mathbf{A}\mathbf{f}(\boldsymbol{\theta}) - \mathbf{y}\|_2^2, \tag{12.6}$$

which yields the network parameters $\hat{\boldsymbol{\theta}}$. With those network parameters, we estimate an image as $\mathbf{f}(\hat{\boldsymbol{\theta}})$.

Coordinate-based networks in its current form do not perform well for denoising, they perform significantly worse than un-trained convolutional networks. But coordinate-based networks perform relatively well for image reconstruction problems arising in MRI, like compressive sensing.

12.5 Proof of Theorem 11

In this section, we prove Theorem 11, for the special case of an MLP without hidden layers. The network has the form

$$f(\mathbf{c}) = \mathbf{w}_1^T \gamma_{\mathbf{W}_0}(\mathbf{c}).$$

With

$$\nabla_{\mathbf{w}_1} f(\mathbf{c}_i) = \gamma_{\mathbf{W}_0}(\mathbf{c}),$$

the (i, j)-th entry of the kernel associated with the network, $\mathbf{K}_{ij}$ is

$$\begin{aligned}\mathbf{K}_{ij} &= \mathbb{E}_{\mathbf{W}_0}\left[\nabla_{\mathbf{w}_1} f(\mathbf{c}_i) \nabla_{\mathbf{w}_1} f(\mathbf{c}_j)\right] \\ &= \mathbb{E}_{\mathbf{W}_0}\left[\gamma_{\mathbf{W}_0}^T(\mathbf{c}_i)\gamma_{\mathbf{W}_0}(\mathbf{c}_j)\right].\end{aligned}$$

Let the entries of $\mathbf{W}$ be $\mathcal{N}(0, \sigma^2)$-distributed, and let $k \to \infty$. Then

$$\gamma_{\mathbf{W}_0}^T(\mathbf{c}_i)\gamma_{\mathbf{W}_0}(\mathbf{c}_j) = e^{-\frac{\sigma^2}{2}\left\|\mathbf{c}_i - \mathbf{c}_j\right\|_2^2}. \tag{12.7}$$

To see this, note that

$$\begin{aligned}\left\langle \gamma_{\mathbf{W}_0}(\mathbf{c}_i), \gamma_{\mathbf{W}_0}(\mathbf{c}_j)\right\rangle &= \frac{1}{k}\sum_{\ell=1}^{k}(\sin(\mathbf{w}_\ell^T \mathbf{c}_i)\sin(\mathbf{w}_\ell^T \mathbf{c}_j) + \cos(\mathbf{w}_\ell^T \mathbf{c}_i)\cos(\mathbf{w}_\ell^T \mathbf{c}_j)) \\ &= \frac{1}{k}\sum_{\ell=1}^{k}\cos(\mathbf{w}_\ell^T (\mathbf{c}_i - \mathbf{c}_j)) \\ &\overset{(i)}{=} \mathbb{E}_{\mathbf{w}}\left[\cos(\mathbf{w}^T(\mathbf{c}_i - \mathbf{c}_j))\right] \\ &= e^{-\frac{\sigma^2}{2}\left\|\mathbf{c}_i - \mathbf{c}_j\right\|_2^2},\end{aligned}$$

where equation (i) holds for $k \to \infty$, and the last equation holds by the characteristic function of a Gaussian random variable. This concludes the proof of Equation (12.7).

The proof generalizes as follows. Consider the a two-layer network, where the first layer is a fixed Fourier-feature map, the second layer is linear with

ReLU-activation, and the third layer is linear. The network is

$$f(\mathbf{x}) = \mathbf{w}_2^T \pi_{\mathbf{W}_1}(\gamma_{\mathbf{W}_0}(\mathbf{c})).$$

Let the entries of $\mathbf{W}_p$ be i.i.d. $\mathcal{N}(0, \sigma_p^2)$-distributed. Define

$$\xi(\rho) = \left(\sqrt{1-\rho^2}/\pi + (1-\cos^{-1}(\rho)/\pi)\rho\right), \quad \zeta(\rho) = (1-\cos^{-1}(\rho)/\pi),$$

and $\phi_\sigma(z) = e^{-\frac{1}{2}\sigma^2 z}$. Then, for $k \to \infty$,

$$\begin{aligned}\mathbf{K}_{ij} &= \nabla_{\mathbf{w}_2} f(\mathbf{c}_i) \nabla_{\mathbf{w}_2} f(\mathbf{c}_j) + \nabla_{\mathbf{W}_1} f(\mathbf{c}_i)^T \nabla_{\mathbf{W}_1} f(\mathbf{c}_j) \\ &= \sigma_1^2 \xi\left(\phi_{\sigma_0}(d_{ij}^2)\right) + \sigma_2^2 \zeta\left(\phi_{\sigma_0}(d_{ij}^2)\right) \cdot \phi_{\sigma_0}(d_{ij}^2).\end{aligned}$$

For networks with more layers, analogous results can be derived so that

$$\mathbf{K}_{ij} = h\left(\exp\left(-\frac{1}{2}\sigma_0^2 \left\|\mathbf{c}_i - \mathbf{c}_j\right\|_2^2\right)\right),$$

where h is a scalar function. This concludes the proof.

12.6 Chapter notes

In this chapter, we discussed the implicit prior of coordinate-based neural networks, and how those networks can be used for signal reconstruction analogously to un-trained neural network. In practice, coordinate-based networks are often used to efficiently represent high-dimensional signals, for example, for scene representations [Mil+20] and also for high-dimensional signals in computational imaging, such as 3D and 4D volumes [KRH24]. A difficulty of working with coordinate-based representations is that they can be computationally expensive, for example, fitting a neural network to represent a scene or to represent another 3D or 4D object can take several GPU hours. To lessen the computational cost, a variety of alternatives have been proposed, for example, signal representations based on hash encodings [Mül+22].

Another interesting alternative to represent high-dimensional signals for computational imaging is Gaussian Splatting [Ker+23], a technique that uses Gaussian functions to represent points in high-dimensional spaces, for example, 3D and 4D spaces.

13
Robustness to perturbations

An important characteristic of reconstruction algorithms beyond reconstruction accuracy is the robustness to a variety of perturbations. In this chapter, we discuss robustness to random and worst-case perturbation in the context of linear inverse problems.

Worst-case perturbations are known as adversarial perturbations in the machine learning literature and are often studied to evaluate the robustness of machine learning systems. For classification problems, there exists a large body of literature on the sensitivity of neural networks to adversarial perturbations [Sze+14; FMF16]. These works show that the predicted class label for a test image can be altered by adding a small perturbation to that image. This observation is attributed to test images lying close to a learned decision boundary, and thus slightly altering the image can cross the boundary from the correct label to an incorrect one.

The study of adversarial robustness for prediction problems, such as classification, is motivated by the concept of an adversary that may alter the input of a machine learning system with imperceptible perturbations. For image reconstruction problems, there is typically no such adversary. However, studying adversarial robustness in image reconstruction methods is important as a variety of perturbations can happen during the measurement process, for example on sensors, pixels can fail, and for magnetic resonance imaging, receiver coils can fail. Studying the worst-case robustness enables to understand and certify the worst-case robustness of imaging systems under such failures.

In this chapter, we will first discuss the worst-case effect a perturbation can have for a simple linear estimator and contrast it to the error caused by a random perturbation. We then discuss the worst-case robustness of a generic estimator and empirical results on the robustness of deep learning-based imaging approaches.

The effect of worst-case perturbations for image reconstruction problems was studied by Cohen et al. [CLH18], Huang et al. [Hua+18], Antun et al. [Ant+20], Darestani et al. [DCH21], and Genzel et al. [GMM22], amongst others. However, it is difficult to study the effect of worst-case perturbations rigorously theoretically or empirically for neural networks, because it is difficult to find worst-case perturbations. To find worst-case perturbations, we typically need to solve a non-convex optimization problem.

Deep Learning for Computational Imaging. Reinhard Heckel, Oxford University Press. © Reinhard Heckel (2025).
DOI: 10.1093/oso/9780198947172.003.0013

In the literature there is some disagreement on whether neural networks or classical methods like ℓ_1-minimization are more sensitive to adversarial perturbations, but it is safe to say that currently there is no evidence that neural networks are fundamentally less stable than classical methods like ℓ_1-regularized least-squares at least for image reconstruction in the context of magnetic resonance imaging [DCH21].

13.1 Worst-case robustness and robustness to random perturbations for a simple estimator

We start with discussing the robustness to a worst-case additive perturbation, potentially selected adversarially. Let $\mathbf{y} = \mathbf{A}\mathbf{x} + \mathbf{e}$ be a noisy measurement of a signal $\mathbf{x}$, where $\mathbf{A} \in \mathbb{R}^{m \times n}$ is a measurement matrix.

We say that a reconstruction algorithm $\Psi\colon \mathbb{R}^m \to \mathbb{R}^n$, for example a neural network, or ℓ_1-regularized least-squares, is stable with respect to a worst-case ℓ_2-perturbation, if for all ϵ-bounded ℓ_2-perturbations the perturbation has a small effect on the reconstructed signal, i.e., the reconstruction error defined as $\|\Psi(\mathbf{A}\mathbf{x} + \mathbf{e}) - \mathbf{x}\|_2$ is small. An ϵ-bounded ℓ_2-perturbation is a perturbation with ℓ_2-norm bounded by ϵ, i.e., it lies in the set $\{\mathbf{e}\colon \|\mathbf{e}\|_2 \leq \epsilon\}$.

To gain intuition on the effect of worst-case perturbations, we consider a simple signal model and reconstruction algorithm that is analytically tractable. Suppose that the signal of interest $\mathbf{x} \in \mathbb{R}^n$ lies in a d-dimensional subspace (with $d \ll n$) spanned by the orthonormal matrix $\mathbf{W} \in \mathbb{R}^{n \times d}$. We consider a reconstruction algorithm that selects the signal that lies in the subspace and is as consistent with the measurement as possible, i.e.,

$$\Psi(\mathbf{y}) = \arg \min_{\mathbf{x}\colon \mathbf{x} = \mathbf{W}\mathbf{c}, \mathbf{c} \in \mathbb{R}^d} \|\mathbf{A}\mathbf{x} - \mathbf{y}\|_2^2.$$

This estimate leverages the prior knowledge that the signal lies in the subspace spanned by $\mathbf{W}$. The estimate is given in closed form as $\Psi(\mathbf{y}) = \mathbf{W}(\mathbf{A}\mathbf{W})^{\dagger}\mathbf{y}$, where $(\cdot)^{\dagger}$ denotes the pseudo-inverse.

As shown below, the reconstruction error induced by a worst-case perturbation is

$$\max_{\mathbf{e}\colon \|\mathbf{e}\|_2 \leq \epsilon} \|\Psi(\mathbf{A}\mathbf{x} + \mathbf{e}) - \mathbf{x}\|_2^2 = \epsilon^2 \frac{1}{\sigma_{\min}^2}, \tag{13.1}$$

where $\sigma_{\min}$ is the smallest singular value of the matrix $\mathbf{A}\mathbf{W}$. Here, we assume that the matrix $\mathbf{A}\mathbf{W}$ has full column rank. Thus, if the smallest singular value of the matrix $\mathbf{A}\mathbf{W}$ is small, a worst-case perturbation has a large effect on reconstruction performance. The deconvolution example from Section 2.3 is an example where the singular value of the involved matrix can be very small.

Let us contrast this to a random perturbation, specifically a random Gaussian perturbation $\mathbf{e} \sim \mathcal{N}(0, \epsilon^2/m\mathbf{I})$. The perturbation has approximately ℓ_2-norm equal to ϵ, i.e., $\|\mathbf{e}\|_2 \approx \epsilon$, just like the worst-case perturbation. To see this note that the expectation of $\|\mathbf{e}\|_2^2$ is $\mathbb{E}\left[\|\mathbf{e}\|_2^2\right] = \epsilon^2$, and $\|\mathbf{e}\|_2^2 = \sum_{i=1}^m e_i^2$ is a sum of independent random variables and concentrates around its expectation.

The effect of the random perturbation is in expectation

$$\mathbb{E}\left[\|\Psi(\mathbf{Ax} + \mathbf{e}) - \mathbf{x}\|_2^2\right] = \epsilon^2 \frac{1}{m} \sum_{i=1}^{d} \frac{1}{\sigma_i^2}, \tag{13.2}$$

where expectation is over the random perturbation $\mathbf{e}$. The proof is at the end of this section.

Thus, the effect of a random, average-case perturbation can be significantly lower than that of a worst-case perturbation: The effect of a random perturbation is determined by the average of the $1/\sigma_i^2$'s, whereas the effect of the worst-case perturbation is determined by the maximum of the $1/\sigma_i^2$'s.

Even if all singular values are equal, and thus the inverse problem is well-conditioned, the effect of a worst-case and a random perturbation can differ significantly. Suppose that all singular values are equal to σ^2. Then the effect of worst-case perturbation is $\epsilon^2 \frac{1}{\sigma^2}$, while that of a random perturbation is $\epsilon^2 \frac{1}{\sigma^2} \frac{d}{m}$. Note that the effect of the random perturbation becomes small as the ratio of the dimension of the signal subspace and the number of measurements, d/m, becomes small. This is because for a random perturbation, only a fraction of d/m of the noise aligns with the measurement-signal subspace (i.e., the subspace spanned by $\mathbf{AW}$), while the worst-case perturbation lies fully in the measurement-signal subspace.

Now let us consider a concrete scenario where the singular values of the matrix $\mathbf{AW}$ are nearly the same, and thus the inverse problem is well-conditioned. The effect of worst- and average-case errors is determined by the singular values of the matrix $\mathbf{AW}$; those singular values depend on both the matrices $\mathbf{A}$ and $\mathbf{W}$. For the case where the measurement matrix $\mathbf{A}$ is random, we can precisely characterize the singular values of $\mathbf{AW}$. Suppose that $\mathbf{A} \in \mathbb{R}^{m \times n}$ has i.i.d. $\mathcal{N}(0, 1/m)$ entries. Then the matrix $\mathbf{AW} \in \mathbb{R}^{n \times d}$ is also a Gaussian matrix with i.i.d. $\mathcal{N}(0, 1/m)$ entries, for any fixed orthonormal matrix $\mathbf{W}$. By the Marchenko-Pastur law (see Rudelson and Vershynin (2010) [RV10] for details), we know that the singular values of the matrix $\mathbf{AW}$ obey asymptotically

$$\sigma_{\min} \to 1 - \sqrt{d/m} \quad \text{and} \quad \sigma_{\max} \to 1 + \sqrt{d/m}.$$

Thus, if $m \gg d$, all singular values are close to one.

To summarize, the effect of worst-case and average-case perturbations can be drastically different, and depends on the measurement matrix and on the interaction of the measurement matrix with the reconstruction algorithm Ψ.

Proof of Equation (13.1) *A worst-case perturbation for the estimator Ψ and for a given signal* **x** *in the d-dimensional subspace is*

$$\hat{\mathbf{e}} = \arg \max_{\mathbf{e}:\|\mathbf{e}\|_2 \leq \epsilon} \|\Psi(\mathbf{Ax} + \mathbf{e}) - \mathbf{x}\|_2^2.$$

With $\mathbf{x} = \mathbf{Wc}$ *for some coefficient vector* **c** *and with* $\mathbf{AW} = \mathbf{U\Sigma V}^T$ *denoting the singular value decomposition of the matrix* **AW**, *we get*

$$\begin{aligned} \|\Psi(\mathbf{Ax} + \mathbf{e}) - \mathbf{x}\|_2^2 &= \left\|\mathbf{W}(\mathbf{AW})^\dagger(\mathbf{AWc} + \mathbf{e}) - \mathbf{Wc}\right\|_2^2 \\ &= \left\|\mathbf{W}(\mathbf{AW})^\dagger \mathbf{e}\right\|_2^2 \\ &= \left\|\mathbf{V\Sigma}^{-1}\mathbf{U}^T\mathbf{e}\right\|_2^2 \\ &= \left\|\mathbf{\Sigma}^{-1}\mathbf{U}^T\mathbf{e}\right\|_2^2. \end{aligned} \tag{13.3}$$

From this, we see that the worst-case perturbation is $\hat{\mathbf{e}} = \epsilon\mathbf{u}_{\text{min}}$, *where* $\mathbf{u}_{\text{min}}$ *is the left-singular vector of* **AW** *corresponding to the smallest singular value. The reconstruction error that this worst-case perturbation induces is given by Equation (13.1).*

Proof of equation (13.2) *From Equation (13.3), we have*

$$\begin{aligned} \mathbb{E}\left[\|\Psi(\mathbf{Ax} + \mathbf{e}) - \mathbf{x}\|_2^2\right] &= \mathbb{E}\left[\left\|\mathbf{\Sigma}^{-1}\mathbf{U}^T\mathbf{e}\right\|_2^2\right] \\ &= \sum_{i=1}^{d} \frac{1}{\sigma_i^2} \mathbb{E}\left[<\mathbf{u}^T{}_i, \mathbf{e}>^2\right] \\ &= \epsilon^2 \frac{1}{m} \sum_{i=1}^{d} \frac{1}{\sigma_i^2}, \end{aligned}$$

which concludes the proof.

13.2 Worst-case robustness of a generic estimator

We next discuss a simple result on the worst-case robustness of a generic estimator or reconstruction algorithm Ψ. The result states that if the measurements corresponding to two different signals $\mathbf{x}, \mathbf{x}'$ that are far from each other are close, and at the same time those signals are reconstructed well by the estimator Ψ, then a relatively small perturbation in the measurement can distort the reconstruction of one of the signals.

Proposition 13 *Suppose there exist two signals* $\mathbf{x}, \mathbf{x}' \in \mathbb{R}^n$ *with corresponding measurements* $\mathbf{Ax}$ *and* $\mathbf{Ax}'$ *that are* ϵ*-close, i.e.,*

$$\|\mathbf{Ax} - \mathbf{Ax}'\|_2 \leq \epsilon. \tag{13.4}$$

Further, suppose that the reconstruction algorithm Ψ *reconstructs both signals* $\mathbf{x}, \mathbf{x}'$ *up to an error of* ϵ*, i.e.,*

$$\|\Psi(\mathbf{Ax}) - \mathbf{x}\|_2 \leq \epsilon, \quad \|\Psi(\mathbf{Ax}') - \mathbf{x}'\|_2 \leq \epsilon. \tag{13.5}$$

Then, there exists an ϵ*-bounded perturbation* $\mathbf{e}$: $\|\mathbf{e}\|_2 \leq \epsilon$ *that can lead the reconstruction algorithm to find a signal close to the alternative signal* $\mathbf{x}'$ *instead of the original signal* $\mathbf{x}$*, i.e.,*

$$\|\Psi(\mathbf{Ax} + \mathbf{e}) - \mathbf{x}'\|_2 \leq \epsilon.$$

The statement has the following interesting implication. Suppose we have two signals that are far (i.e., $\mathbf{x} - \mathbf{x}'$ is large), but the corresponding measurements $\mathbf{Ax}$ and $\mathbf{Ax}'$ are close. Then a small perturbation $\mathbf{e}$ can have a large effect on a reconstruction algorithm that works well, in that the reconstruction algorithm returns a vector close to $\mathbf{x}'$ as opposed to returning a vector close to $\mathbf{x}$ as desired. Note that for this scenario, two conditions have to be satisfied: (i) the differences between measurements, $\|\mathbf{Ax} - \mathbf{Ax}'\|_2$, is small while the difference between signals, $\mathbf{x} - \mathbf{x}'$, is large (i.e., the difference $\mathbf{x} - \mathbf{x}'$ lies close to the null-space of the matrix $\mathbf{A}$), and (ii) the reconstruction algorithm has to work well on both signals (i.e., Condition (13.5) has to be satisfied). This is illustrated in Figure 13.1.

This result illustrates the interplay of the forward map and the reconstruction algorithm in determining the robustness. For example, suppose we have a forward map $\mathbf{A}$ that maps two signals $\mathbf{x}, \mathbf{x}'$ with a large distance to measurements $\mathbf{Ax}$ and $\mathbf{Ax}'$ that are close.

An algorithm can stably reconstruct signal $\mathbf{x}$ (i.e., reconstruct a signal close to $\mathbf{x}$ under a perturbation) if it does not do well at reconstructing the signal $\mathbf{x}'$, for example by excluding the signal $\mathbf{x}'$ from the signals that can be reconstructed. An example is an algorithm, like ℓ_1-regularized least-squares, that stably reconstructs a sparse signal, but fails to reconstruct a dense signal. In addition, for stably reconstructing a signal, signals that are far should also lead to measurements that are far in the measurement space. For sparse signals, we saw earlier that taking measurements with a random Gaussian matrix yields measurements that are well-separated in the measurement space, specifically, the distance between pairs of sparse signals is around the distance between the corresponding measurements. If this property is satisfied, it can be shown that reconstruction of sparse signals with ℓ_1 minimization is stable.

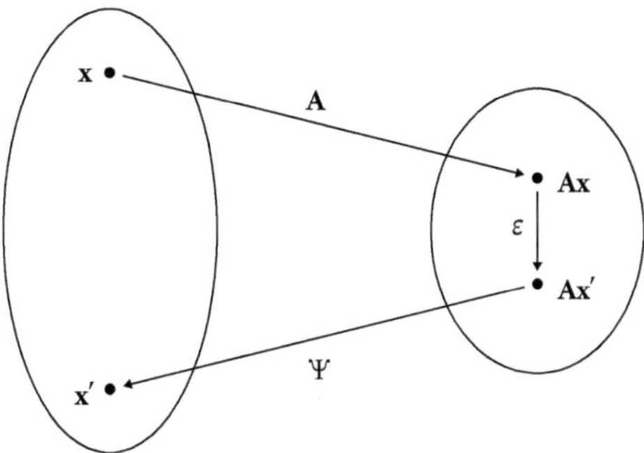

Figure 13.1 An illustration of how a perturbation can lead to a large error. Suppose we take a measurement **Ax** of a signal **x**, and add a perturbation to it. This moves the measurement to a measurement of another signal **x**′, equal to **Ax**′. If the reconstruction algorithm Ψ succeeds at reconstructing **x**′, it will yield the signal **x**′. If the signal **x**′ is far from the signal **x**, then the error induced by the perturbation is large.

Proof of Proposition 13: Let $\mathbf{e} = \mathbf{A}(\mathbf{x}' - \mathbf{x})$, and note that from the measurements of the two signals being ϵ-close (i.e., from (13.4)), we have that $\|\mathbf{e}\|_2 \leq \epsilon$. Moreover,

$$\|\Psi(\mathbf{Ax} + \mathbf{e}) - \mathbf{x}'\|_2 = \|\Psi(\mathbf{Ax} + \mathbf{A}(\mathbf{x}' - \mathbf{x})) - \mathbf{x}'\|_2 = \|\Psi(\mathbf{Ax}') - \mathbf{x}'\|_2 \leq \epsilon,$$

where the inequality holds by assumption (13.5). This concludes the proof. □

13.3 Empirical robustness in the context of accelerated MRI

In this section we discuss an empirical result from Darestani et al. (2021) [DCH21] on the effect of small adversarial perturbations on the robustness of three different classes of reconstruction algorithms.

The experimental setup is as follows. Ten randomly-chosen (proton-density-weighted) knee MRI images from the fastMRI validation set are considered, and for each image an ϵ-bounded perturbation is added to the measurement (k-space). The ϵ-bounded perturbation is generated for each image and optimization algorithm Ψ (U-net, ℓ_1-regularized least-squares, and an un-trained neural network, specifically a version of the deep decoder discussed earlier) by solving the optimization problem

$$\hat{\mathbf{e}} = \arg \max_{\{\mathbf{e}:\|\mathbf{e}\|_2 \leq \epsilon\}} \|\Psi(\mathbf{Ax} + \mathbf{e}) - \Psi(\mathbf{Ax})\|_2.$$

This optimization problem is non-convex. We generate a perturbation $\hat{\mathbf{e}}$ with the projected gradient method that alternates between a gradient step and projection

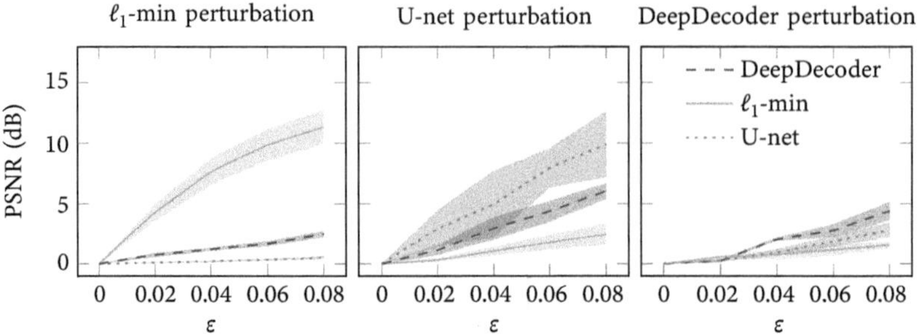

Figure 13.2 Both trained and un-trained reconstruction methods are similarly vulnerable to small adversarial perturbations. Performance loss as a function of the perturbation strength, $\epsilon = \frac{\|\text{perturbation}\|_2}{\|k\text{-space}\|_2}$, for all methods. In each plot, the perturbations are obtained by attacking one method (specified in the plot title), and are applied to all methods. The results are averaged over 10 randomly-chosen proton density knee images from the fastMRI validation set. Shaded areas denote 95% confidence intervals.

onto the set of admissible perturbations $\{\mathbf{e}: \|\mathbf{e}\|_2 \leq \epsilon\}$. Then, each perturbation is applied to each image, and reconstructed with all three methods. Figure 13.2 shows the results, and Figure 13.3 shows a few example images after reconstruction under an adversarial perturbation.

It can be seen that both trained (U-net) and un-trained methods (ℓ_1-regularized least-squares and the deep decoder) are sensitive to small adversarial perturbations. The experiment demonstrates that while adversarially-selected perturbations for a trained network (e.g., U-net) have a relatively mild impact on ℓ_1-norm minimization, the converse is also true: perturbations found for ℓ_1-norm minimization have a significant effect for ℓ_1-minimization, but only a mild effect on U-net's performance. Thus, both methods are (similarly) vulnerable to perturbations specifically tailored to them.

At first sight, the findings in Figure 13.2 suggest that deep decoder is slightly more robust than both U-net and ℓ_1-based reconstruction. However, we cannot draw such a conclusion, simply because in order to find adversarial perturbations, we solve a non-convex optimization problem with a numerical method (projected gradient descent), and we are not guaranteed to find a worst-case perturbation. In principle, it could be that for ℓ_1-minimization we find a worst-case perturbation but for U-net we do not, for example. This issue is inherent to the problem setup and applies to all current methods for finding adversarial perturbations.

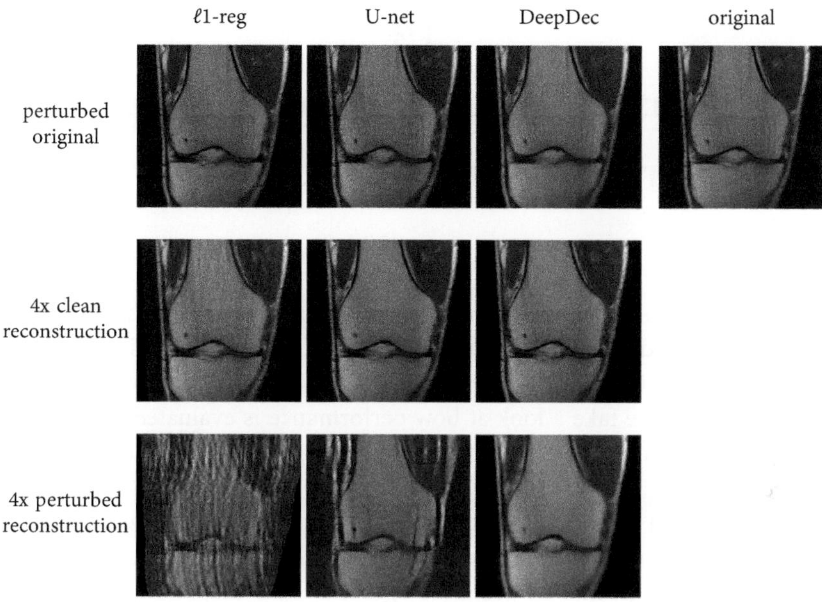

Figure 13.3 Example images with small adversarially-selected perturbations as well as the reconstructions of different reconstruction algorithms considered in this book. The first row shows the original image and perturbed original images after finding a perturbation specifically for each method, with $\epsilon = 0.08$. The second row shows reconstructions from the clean 4x under-sampled measurements of the reference image. The third row shows the reconstructions from the perturbed 4x under-sampled measurements.

14
Datasets and evaluation of image reconstruction methods

Image and measurement data is a critical component for training and evaluating deep learning-based image reconstruction methods. Data is also critical for tuning and evaluating traditional, non-learning-based image reconstruction methods.

In this chapter, we take a look at how performance is evaluated, at the training and evaluation data for image reconstruction algorithms, and how the data is obtained.

For training and evaluating image reconstruction algorithms, we ideally have measurements (for example a noisy image, or an MRI measurement) and corresponding target images that are the desired ground-truth images. Neural networks trained in a supervised fashion require such data for training. But even approaches that do not rely on supervised training, like generative model-based approaches that only require clean images for training, typically require pairs of measurement and target image for evaluation.

However, in most imaging applications, it is difficult or impossible to collect pairs of measurements and corresponding ground-truth images. For example, suppose we wish to collect data for denoising camera images. It is easy to collect noisy images with a camera by taking pictures, but it is difficult to get matching noise-free images, since the images collected with a camera are inherently noisy. Similarly, it is easy to collect undersampled MRI measurements or low-dose CT measurements but it is difficult to obtain associated target images for training. In practice, we might collect more MRI measurements or perform a high-dose CT scan and reconstruct a target image from those, but the resulting image is not noise-free or clean.

Training and evaluation data can, in principle, be synthesized from a suitable dataset of clean images, for example, from photographs from the Internet, if we are interested in training an algorithm for denoising natural images. This can work well, in particular if we can simulate the forward model well. However, this can also yield to misleading and biased results as we will discuss in Section 14.2.

Ideally, the training and test data is reflective of the data the algorithm is ultimately applied to. For example, if we use a method for accelerated MRI to reconstruct knees in a given hospital, we ideally train the method on data that comes from the same distribution. In practice, this is very difficult, and we typically have so-called distribution shifts between the training and test data. For

Deep Learning for Computational Imaging. Reinhard Heckel, Oxford University Press. © Reinhard Heckel (2025).
DOI: 10.1093/oso/9780198947172.003.0014

example, we might train and test a method for accelerated MRI to reconstruct knees on a dataset of knees that we collected at a hospital, but ultimately we use the method in another hospital. The patient population and scanner setup in the new hospital is slightly different, which affects the performance of the algorithm. In Section 14.4 we discuss the effect of such distribution shifts on the performance of algorithms.

Typically, imaging performance is measured by comparing a reconstructed image and the ground-truth image with a similarity metric. It turns out that measuring the similarity of two images in a way that correlates with human perception is very difficult, which we discuss in Section 14.3.

We close this chapter with remarks on biases and hallucinations.

14.1 A tour of datasets

We start with looking at a few influential datasets in imaging and discuss how they are used in image reconstruction research, to illustrate common practices and challenges.

14.1.1 Standard test images

The performance of image reconstruction methods can be evaluated on a small set of example images. Since traditional image reconstruction algorithms are not learning-based and thus require no training data; traditionally, image datasets for tuning and evaluating image reconstruction algorithms are relatively small.

A small set of digitalized test images are used in tens of thousands of publications since the 1970s. Those images include a 1978 greyscale image of a cameraman at MIT, the Lena image that was in use since 1973, and a variety of images from the USC-SIPI image database. The USC-SIPI is a set of digitalized images first distributed in 1977 that includes popular test images like the pirate, boat, airplane, and peppers images displayed in Figure 14.1. Those test images contain features that make them useful for evaluating image reconstruction methods, such as light and dark regions, flat regions, smoothly varying regions, sharp edges, and textured regions.

14.1.2 Image datasets crawled from the web

Imaging algorithms are often trained and evaluated on image datasets crawled from the web, often originally for purposes other than image reconstruction, such

Figure 14.1 Four popular test images from the USC-SIPI image database that are used since the 1970s as test images to illustrate image processing algorithms, such as denoising algorithms.

as object recognition and classification. For example, several denoising algorithms (e.g., DnCNN by Zhang et al. [Zha+17]) are trained and evaluated on part of the Berkeley segmentation dataset (BSD68) [Mar+01], a dataset originally collected for image segmentation.

Another popular dataset is the ImageNet [Den+09], a dataset consisting of about 14 million labelled images containing a variety of objects and scenes. ImageNet was collected for image recognition, but a variety of image reconstruction methods are trained and evaluated on ImageNet, for example the super-resolution method by Dong et al. [Don+16]. The ImageNet images and images from similar datasets are taken from public image hosting sites, such as Flickr.

Interestingly, the vast majority of image datasets crawled from the web and used for training and evaluating image reconstruction algorithms is not collected for that purpose specifically, but for other tasks such as image classification or segmentation. There are a few exceptions, for example Agustsson and Timofte [AT17] built a super-resolution challenge dataset called DIV2K (this stands for diverse 2K) by manually collecting high-resolution images from the Internet, and paying special attention to image quality and contents and diversity of sources.

Image datasets crawled from the web only consist of images. To construct datasets for training and/or evaluation of image reconstruction methods, measurements are generated from the images. For example, for Gaussian denoising, measurements are obtained by adding zero-mean Gaussian noise, and for super-resolution, the images are downsampled with bicubic downsampling.

14.1.3 Image denoising datasets

Image reconstruction algorithms are often trained and evaluated on measurements from synthesized images since pairs of measurements and ground-truth images are very difficult to collect. This can be suboptimal when the synthesized measurements do not reflect true measurements.

For image denoising, it is possible to obtain pairs of measurements and corresponding target images by taking multiple measurements of a static scene. One of those measurements can serve as the noisy measurement, and from the others it is possible to estimate a ground-truth image, for example, through averaging the noisy measurements, and this estimate can serve as a target for training and evaluation. While from many noisy images we can get a target image that is close to a ground-truth, such an estimate still contains noise.

The RENOIR dataset is an example of such a dataset [AB18]. The RENOIR dataset consists of colour images of 120 scenes naturally corrupted by low-light noise. A low-noise image was obtained with low light sensitivity and long exposure time, this is an estimate of the ground-truth image and taken as the target image, and corresponding noisy images (measurements) were collected with increased light sensitivity and reduced exposure time.

The smartphone image denoising dataset (SIDD) [ALB18] is another example of such a real-noise dataset. The SIDD dataset consists of images from 10 scenes taken under different lighting conditions with different smartphone cameras and corresponding estimates of the ground-truth images. For each static scene and lighting condition, 150 images were sequentially acquired and combined through correction and alignment steps to a target image. Each of the individual images can serve as a noisy measurement of the ground-truth image. See Figure 14.2 for an example of a noisy image along with the target/estimated ground-truth image.

In all those cases, the target image is an estimate of the ground-truth, and how close this estimate is to a clean ground-truth image depends on how good those assumptions are, and how good the estimate is.

Figure 14.2 A crop of an image along with the estimated target image from the SIDD dataset by Abdelhamed et al. [ALB18] licensed under the MIT License.

14.1.4 fastMRI

fastMRI [Zbo+19] is a large dataset of fully-sampled MRI measurements for training and evaluating algorithms for accelerated 2D MRI. Fully-sampled means the entire k-space has been collected. The first two releases of the dataset contain data of scans of knees and brains, and recently prostate and breast scans have been released too. The dataset is the largest publicly available dataset for accelerated MRI research. The knee and brain datasets consist of 1200 and 6400 volumes, which results in 42k and 100k 2D slices in total. Target images for training and evaluation can be computed by reconstructing an image (a slice in the volume) based on the entire data of that slice. Corresponding undersampled measurements for training and evaluation can be obtained by retrospectively subsampling the raw data.

14.2 Inverse crimes

Since ground-truth images corresponding to measurements are fundamentally difficult to acquire, it is common to train and evaluate neural networks (and image reconstruction algorithms more broadly) on data generated by simulating the measurement generation, instead of relying on actual physical measurements. Concrete examples are super-resolution based on downsampling high-resolution images, denoising based on adding Gaussian noise to measurements, and accelerated MRI reconstruction based on generating measurements from given images using the known MRI forward model.

When simulating the measurement generation process, it is possible to simplify the image reconstruction problem to an extent that makes solving the inverse problem of reconstructing the image relatively easy and the associated results misleading. Often the results are overly optimistic since the synthetic data perfectly matches the assumptions of the reconstruction algorithm, which is not realistic in practice where complexities of the real data are not accounted for in the model. This trap is called an inverse crime, and is described in the classical literature on inverse problems [CK19], [KS05, Sec. 5.8].

14.2.1 Performance relations in denoising

It is very common to evaluate denoising algorithms on Gaussian denoising, even though we might be interested in the performance of algorithms evaluated on real-world noise. Algorithms trained and evaluated on real-world noise perform better or worse than algorithms trained and evaluated on Gaussian noise, as we will see in this section.

Thus training and evaluating on Gaussian noise can, in general, not serve as a proxy for training and evaluating on real-world noise. However, for the scenario discussed here, we find that the performance for Gaussian noise and real-world noise is strongly correlated. Thus, if we are interested in relative performance comparisons of reconstruction algorithms (which we are often in research and for benchmarking), training and evaluating on Gaussian noise might be sufficient.

Let us consider the two data processing pipelines illustrated in Figure 14.3.

Real-world camera noise (SIDD). We consider the smartphone image denoising dataset (SIDD) [ALB18] mentioned earlier, consisting of images from 10 scenes taken under different lighting conditions with different smartphone cameras and corresponding estimated ground-truth images. The SIDD Medium Dataset is taken for training, and the SIDD validation set for testing.

Gaussian noise (synthesized data). We take ImageNet images as ground-truth training images and the Kodak24 test images as ground-truth images for testing. To obtain associated noisy measurements, the images are contaminated with zero-mean Gaussian noise with variance $\sigma^2 = 25^2$ (the pixel range is 0–255).

Real-world camera noise

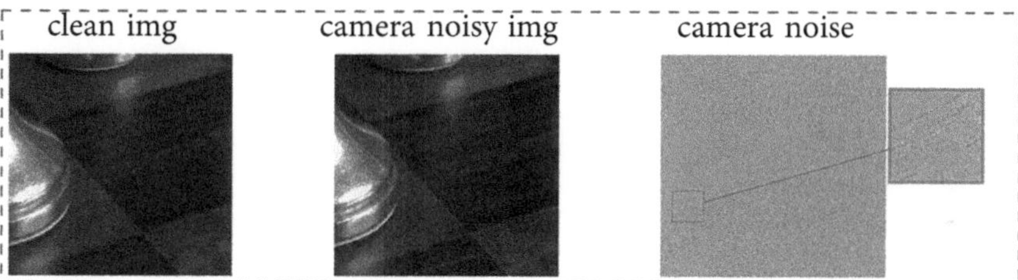

Gaussian noise (synthesized data)

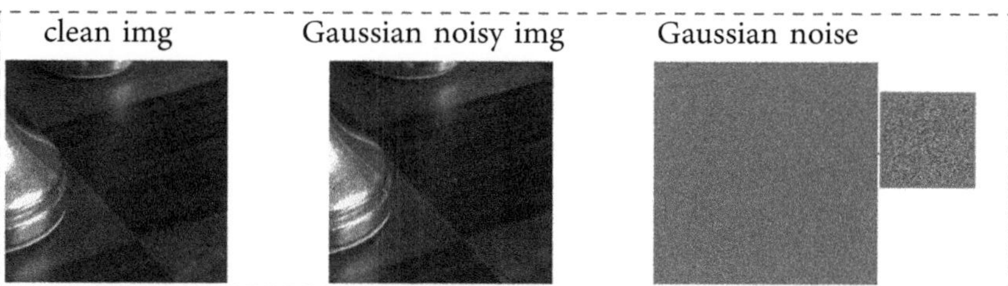

Figure 14.3 Two data processing pipelines for denoising. For the real-world camera noise, an image is collected through long exposure at low ISO, and the noisy image corresponds to taking the same image with short exposure and high ISO. The noise is signal dependent and non-Gaussian. For the synthesized data, Gaussian noise is added to an image. Both the camera noisy image and the Gaussian noisy image have 23.7dB PSNR.

In Figure 14.4, we see the performance of a variety of algorithms trained and evaluated on the Gaussian denoising data as a function of the performance of the respective algorithm trained and evaluated on the real-world camera noise (SIDD) data. The noise level (measured in PSNR) for Gaussian and real-world noise is the same. The algorithms are a variety of networks trained end-to-end (U-net [RFB15], a standard baseline method; DnCNN [Zha+17], a simple CNN which is also considered a baseline method; Restormer [Zam+22] and SwinIR [Lia+21] which both use elements from transformers and are considered state-of-the-art). Each network is trained on {250,500,1000,2000,4000} images to get a large variety of models (networks trained on more data perform better). The algorithms also include the classical BM3D [Dab+07] algorithm designed for Gaussian denoising. BM3D is tailored for Gaussian noise, the algorithm design assumes Gaussian noise.

The plot shows that the networks perform better on real-world denoising than Gaussian denoising, i.e., scores are achieved. This demonstrates that we cannot use Gaussian denoising as a proxy for real-world denoising, as expected. However, the plot also shows that the performance of the neural networks trained and evaluated on the Gaussian data and the real-world noise are well correlated. This indicates that the ranking of architectures is preserved when evaluated on Gaussian and natural noise, even though the noise is quite different.

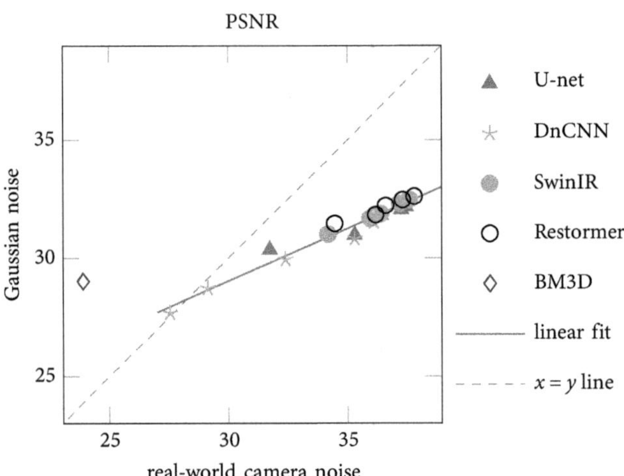

Figure 14.4 Denoising performance in dB for four denoising networks, each trained and evaluated on Gaussian noise, and on real-world camera noise. The networks perform better on real-world camera noise than on Gaussian noise, but the ranking of the algorithms on each of the datasets is approximately preserved.

BM3D performs relatively better on Gaussian noise than on natural noise, which is not surprising since it is designed for Gaussian noise. This demonstrates that comparisons amongst algorithms can have different results when compared on synthetic versus real data.

14.2.2 Performance relations in MRI

In a recent paper, Shimron et al. [Shi+22] highlighted two concrete scenarios where misuse of data based on simulating a measurement process leads to biased performance results for accelerated MRI. MRI measurements are acquired in the Fourier domain. To accelerate MRI, only a subset of the measurements is acquired. For developing and evaluating deep learning algorithms for MRI, ideally, fully-sampled measurements are available. Shimron et al. [Shi+22] notes that since only few such datasets are available, and the existing datasets are sometimes difficult to handle, it is common that researchers synthesize measurements based on processed MRI images. Shimron et al. [Shi+22] found that on the processed data, the image reconstruction algorithms considered perform significantly better than on the original data, similar as in denoising example that we saw in the previous section.

Similarly, as for the denoising data, however, the performance of algorithms trained and evaluated on Shimron et al.'s [Shi+22] synthesized data is strongly linearly correlated with the algorithms trained and evaluated on the real data. Thus, relative performance comparisons made by training and evaluating on synthesized data often remain valid, and the common practice of synthesizing data for evaluating image reconstruction algorithms is often useful for distinguishing the performance of algorithms.

However, this is not generally the case and care must be taken when training and evaluating on synthesized data.

14.3 Measuring the similarity of two images

It is difficult to capture the similarity between a ground-truth image and a reconstructed image with a single number. We start with discussing the popular mean-squared error along with shortcomings for measuring the perceptual image quality. Next, we discuss the structural similarity index metric, a better measure for the perceived image quality, and metrics based on features of neural networks, which even better correlate with human judgement. Finally, we close with comments on other similarity metrics between images and general difficulties in measuring similarity.

14.3.1 Mean-squared error

The arguably most popular quantitative performance metric in signal and image processing is the mean-squared error (MSE). It is used to assess image quality, for comparison of competing methods, and it is also often used as a loss function for training deep neural networks for image reconstruction. The MSE is popular for imaging applications, despite the fact that the MSE is often criticized for not being a good proxy for measuring the perceptual quality of image reconstructions or the perceptual similarity of two images.

The MSE between two signals $\mathbf{x}, \mathbf{x}' \in \mathbb{R}^n$ is defined as

$$\mathrm{MSE}(\mathbf{x}, \mathbf{x}') = \frac{1}{n}\|\mathbf{x} - \mathbf{x}'\|_2^2.$$

The MSE is often expressed as the Peak Signal-to-Noise-Ratio (PSNR) defined as

$$\mathrm{PSNR} = 10 \log_{10} \frac{R^2}{\mathrm{MSE}},$$

where R is the dynamic range of the signal. For example, $R = 256$ if we have greyscale values in the range $\{0, 1, ..., 255\}$, and $R = 1$ if the pixels are in the interval $[0, 1]$. The MSE is popular for the following reasons:

- The MSE is simple to compute and simple to differentiate, and therefore also simple to optimize over.
- The MSE is proportional to a metric or distance function, i.e., the ℓ_2-norm. A metric d satisfies the following conditions: (i) Identity: $d(\mathbf{x}, \mathbf{x}') = 0$ if and only if $\mathbf{x} = \mathbf{x}'$. (ii) Symmetry: $d(\mathbf{x}, \mathbf{x}') = d(\mathbf{x}', \mathbf{x})$. (iii) Triangle inequality: $d(\mathbf{x}, \mathbf{x}') \leq d(\mathbf{x}, \mathbf{z}) + d(\mathbf{z}, \mathbf{x}')$.
- The MSE has a physical meaning in that it is a natural way to define the energy of the error signal. The energy of the signal is preserved under orthonormal or unitary transformations, such as the Fourier transform.

While those are all useful and favourable properties, the MSE does not measure the perceived image quality well. A trivial example is a slight mean shift. Adding a constant to the signal $\mathbf{x}' = \mathbf{x} + c$ changes the MSE relative to $\mathbf{x}$, but does not change any of the details of the image. Other examples of perturbations that lead to a similar MSE yet to varying levels of perceived image quality can be found in Figure 14.5, see also Wang and Bovik [WB09, Fig. 1] and Zhou et al. [Zho+04, Fig. 2] for more examples.

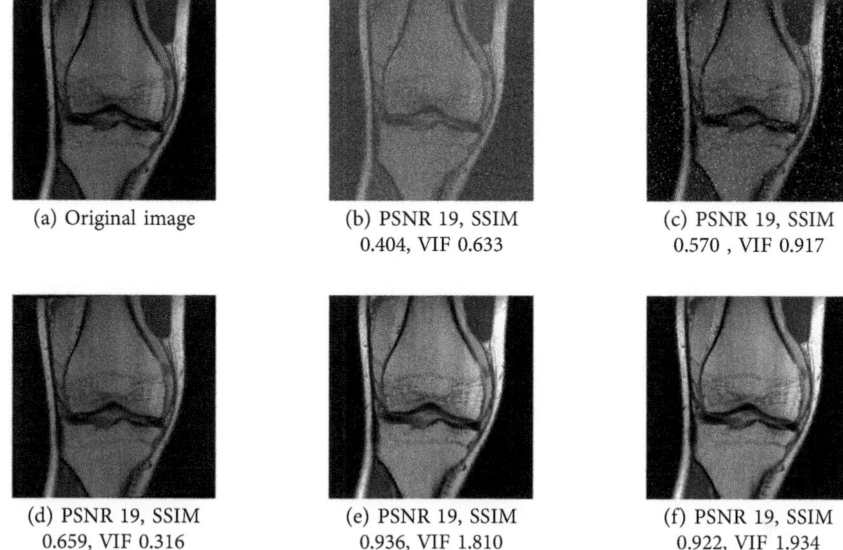

(a) Original image

(b) PSNR 19, SSIM 0.404, VIF 0.633

(c) PSNR 19, SSIM 0.570 , VIF 0.917

(d) PSNR 19, SSIM 0.659, VIF 0.316

(e) PSNR 19, SSIM 0.936, VIF 1.810

(f) PSNR 19, SSIM 0.922, VIF 1.934

Figure 14.5 Comparison of image assessment metrics for a knee image from the fastMRI dataset perturbed with different distortions. (a) Ground-truth image. (b) Gaussian noise contamination. (c) Salt and pepper noise. (d) Slight anti-clockwise rotation. (e) Luminance shift. (f) Contrast shift. While all images have the same PSNR, the ranking in terms of SSIMs and VIF are better proxies for the perceived similarity to the original image. However, the ranking in terms of SSIM and VIF is not consistent either.

14.3.2 Structural similarity index measure (SSIM)

The SSIM is a measure for the perceived similarity of two images. It was introduced by the 2004 paper [Zho+04] and is very popular because the SSIM provides a better metric for comparing the perceptual quality of two images and at the same time the SSIM retains a few desirable properties of the MSE: SSIM is also simple to compute and is differentiable. The latter is particularly important for deep learning-based reconstruction methods, because we can use the SSIM as a loss function to train neural networks.

The SSIM aims to measure "perceived changes in structural information" [Zho+04]. Let $\mathbf{x}$ and $\mathbf{x}'$ be two non-negative local image patches extracted from an image. The local SSIM relies on three comparisons: luminance (brightness value) $\ell(\mathbf{x}, \mathbf{x}')$, contrast $c(\mathbf{x}, \mathbf{x}')$, and structure $s(\mathbf{x}, \mathbf{x}')$. Those three similarity measures are defined as functions of the mean and the standard variance of the patches, and are combined to an overall similarity measure as

$$\mathrm{SSIM}(\mathbf{x}, \mathbf{x}') = \ell(\mathbf{x}, \mathbf{x}') \cdot c(\mathbf{x}, \mathbf{x}') \cdot s(\mathbf{x}, \mathbf{x}'). \tag{14.1}$$

Each of the functions returns a value of 1 if the signal are most similar in the respective comparison measure, and are smaller otherwise. The three functions could also be combined differently than in the SSIM definition (14.1) to a single metric, to account for their relative importance. The choice to weigh them equally was made for simplicity.

The luminance comparison is defined as

$$\ell(\mathbf{x}, \mathbf{x}') = \frac{2\mu_x \mu_{x'} + \epsilon_\ell}{\mu_x^2 + \mu_{x'}^2 + \epsilon_\ell},$$

where the mean intensity of the image patch $\mathbf{x} \in \mathbb{R}^n$ is defined as

$$\mu_x = \frac{1}{n} \sum_{i=1}^{n} x_i.$$

Moreover, ϵ_ℓ is a small numerical constant, added for numerical stability to deal with a pair of signals $\mathbf{x}, \mathbf{x}'$, where $\mu_x^2 + \mu_{x'}^2$ is very small. The luminance function is $\ell(\mathbf{x}, \mathbf{x}') = 1$ if both signals $\mathbf{x}$ and $\mathbf{x}'$ have the same mean intensity, and it is small if both have very different mean intensities. Note that we can have $\ell(\mathbf{x}, \mathbf{x}') = 1$ even when $\mathbf{x} \neq \mathbf{x}'$.

The contrast comparison is defined as

$$c(\mathbf{x}, \mathbf{x}') = \frac{2\sigma_x \sigma_{x'} + \epsilon_c}{\sigma_x^2 + \sigma_{x'}^2 + \epsilon_c},$$

where again ϵ_c is a constant added for numerical stability, and the standard deviation of the signal $\mathbf{x} \in \mathbb{R}^n$ is

$$\sigma_x = \sqrt{\frac{1}{n-1} \sum_{i=1}^{n} (x_i - \mu_x)^2}.$$

Again, if the two signals have the same contrast, the contrast comparison function is $c(\mathbf{x}, \mathbf{x}') = 1$, and if they have very different contrast, the contrast comparison is small; but again $c(\mathbf{x}, \mathbf{x}') = 1$ does not imply that $\mathbf{x} = \mathbf{x}'$. The contrast comparison function measures the variation of the signal.

Finally, the structural comparison is defined as the correlation (inner product) of the two signals after subtracting the mean and normalizing with the standard deviation

$$s(\mathbf{x}, \mathbf{x}') = \frac{\sigma_{xx'} + \epsilon_s}{\sigma_x \sigma_{x'} + \epsilon_s}, \tag{14.2}$$

where the covariance of the two variables $\mathbf{x}, \mathbf{x}'$ is defined as

$$\sigma_{xx'} = \frac{1}{n-1} \sum_{i=1}^{n} (x_i - \mu_x)(x'_i - \mu_{x'}).$$

The structural comparison again has the property that $s(\mathbf{x}, \mathbf{x}') = 1$ does not imply that $\mathbf{x} = \mathbf{x}'$.

Putting all together yields the definition of the SSIM as

$$\begin{aligned} \mathrm{SSIM}(\mathbf{x}, \mathbf{x}') &= \frac{2\mu_x \mu_{x'} + \epsilon_\ell}{\mu_x^2 + \mu_{x'}^2 + \epsilon_\ell} \frac{2\sigma_x \sigma_{x'} + \epsilon_c}{\sigma_x^2 + \sigma_{x'}^2 + \epsilon_c} \frac{\sigma_{xx'} + \epsilon_s}{\sigma_x \sigma_{x'} + \epsilon_s} \\ &= \frac{(2\mu_x \mu_{x'} + \epsilon_\ell)(2\sigma_{xx'} + \epsilon_c)}{(\mu_x^2 + \mu_{x'}^2 + \epsilon_\ell)(\sigma_x^2 + \sigma_{x'}^2 + \epsilon_c)}, \end{aligned}$$

where the last equality follows from setting $\epsilon_s = \epsilon_c/2$. The SSIM satisfies the following desirable conditions: It is symmetric, i.e., $\mathrm{SSIM}(\mathbf{x}, \mathbf{x}') = \mathrm{SSIM}(\mathbf{x}, \mathbf{x}')$, it is bounded by one, i.e., $\mathrm{SSIM}(\mathbf{x}, \mathbf{x}') \leq 1$, and it has a unique maximum $\mathrm{SSIM}(\mathbf{x}, \mathbf{x}') = 1$ if and only if $\mathbf{x} = \mathbf{x}'$.

The SSIM is applied locally to image patches rather than globally. To account for edge or blocking artifacts, the pixels in each block are typically weighted by multiplying them with the Gaussian weighting function that slightly decays from the centre of the blocks. The overall SSIM of the images $\mathbf{x}, \mathbf{x}' \in \mathbb{R}^n$ is then the average of the local SSIMs, with the average taken over the image patches $(\mathbf{x}_i, \mathbf{x}'_i)$:

$$\mathrm{SSIM}(\mathbf{x}, \mathbf{x}') = \frac{1}{M} \sum_{i=1}^{M} \mathrm{SSIM}(\mathbf{x}_i, \mathbf{x}'_i). \tag{14.3}$$

The SSIM is applied locally and aggregated to a global SSIM value for the following reason. Images often contain varying structures and textures at different locations, and evaluating SSIM locally enables to capture those local variations in luminance, contrast, and structure. The human visual system is more sensitive to local changes in image quality rather than global changes. By computing the SSIM locally and summing up the results aligns better with human perception.

There are several refinements of the SSIM, and one of the best-performing (as measured by human studies of perceptive quality) is the multiscale SSIM proposed by [WSB03], which applies the SSIM to different low-pass filtered and downsampled versions of the image, to measure similarity at different resolutions.

14.3.3 Metrics based on features of neural networks

As we saw above, PSNR and SSIM are often not consistent with the human perception on how close two images are. For example, blurred reconstructions can result in large PSNRs relative to the original image, but cause a large perceptual loss. See Figure 14.5 for a few examples of perturbations and associated PSNRs and SSIMs.

In search of a perceptual metric that better aligns with human evaluation, Zhang et al. [Zha+18] considered the distance between features of neural networks trained on a perception task, specifically image classification. Feature distance metrics have been used as a loss for training neural networks in prior works.

Let ψ be a neural network that maps an image to a feature space. A deep feature metric based on the neural network ψ is

$$\delta(\mathbf{x}, \mathbf{x}') = \|\psi(\mathbf{x}) - \psi(\mathbf{x}')\|_2^2.$$

Here, the metric used to compute the distance in feature space is the ℓ_2-distance, but other metrics to compute the distance in feature space are also possible.

A variety of choices of neural networks work well. Zhang et al. [Zha+18] considered standard convolutional networks, such as VGG-16 or AlexNet. Those networks consist of a few convolutional layers that generate a feature representation of the image followed by a few fully-connected layers that map the feature representation to a prediction of a class label. The networks were trained on the ImageNet dataset, discussed earlier, to perform classification. It is also possible to use image encoders trained in an un-supervised fashion, such as the encoder in Contrastive Language-Image-Pre-training (CLIP) [Rad+21].

Zhang et al. [Zha+18] evaluated such deep feature metrics on human judgements and found that deep feature metrics based on networks trained in a supervised fashion as well as trained in an un-supervised fashion correlate well with human evaluations, significantly better than classical metrics like PSNR and SSIM do. Training the network is critical. Simply using a randomly initialized, un-trained network to generate the feature representation does not result in a good perceptual metric. Thus, networks trained to solve visual prediction and modelling tasks (such as classification, or alignment with language) learn a representation that enables to measure perceptual distance well. Complementary, models trained in a self-supervised fashion, for example on image-text pairs (CLIP), or by masking image patches, enable to learn representations that work well for prediction tasks when fine-tuned.

The visual perception as evaluated by humans simply measures how close a human believes two images are, yet there might be small differences that can be important in applications, for example pathologies in medical images, that are not perceived well by non-experts. However, Adamson et al. [Ada+23] evaluated deep feature metrics for medial imaging, specifically for MRI, and found that deep feature metrics correlate well with the evaluation of (expert) radiologists. This holds true, even if the network ψ used to generate the feature representation is not trained on MRI images, but on natural images such as those from ImageNet instead.

Combined, those results suggest that deep feature-based metrics are a powerful choice for measuring the reconstruction accuracy of image reconstruction methods. In current research papers and benchmarks, however, deep feature-based metrics are rarely used which is perhaps because the community is not yet used to them and also the value of the metric depends on the neural network used, thus there is no single metric that people are used to.

14.3.4 Assessing image quality is difficult

In this section, we discussed three metrics for measuring the similarity between two images in order to assess the performance image reconstruction algorithms.

The first, PSNR, is based on measuring the difference at a pixel level. The second, SSIM, is often categorized as a perceptual metric because it measures the image quality based on structural information, luminance, and contrast. There are other notions of perceptual metrics, such as the visual information fidelity [SB06] (VIF), not discussed in detail here, which evaluates the amount of visual information preserved relative to a reference image.

Third, we discussed deep feature metrics which are based on learning, contrary to pixel wise (PSNR) and perceptual metrics (SSIM and VIF).

All those metrics have in common that they capture the image quality in a single number. However, as became clear when examining the different metrics, the image quality is difficult to measure and to capture in a single number.

Which metric is best depends on the application and a variety of considerations. For example, suppose we are interested in evaluating the quality of an MRI image. In this application, we are primarily interested in the image retaining its diagnostic value, i.e., whether the diagnosis of a radiologist based on the original image and the diagnosis of a radiologist based on the reconstructed image are the same.

To measure this, Mason et al. [Mas+20] compared the image quality scores of popular metrics including MSE, SSIM, and VIF to the perception of the diagnostic value of five radiologists. The study found that VIF correlates better with the radiologists opinions than the more frequently used metrics SSIM and MSE. As mentioned before, Adamson et al. [Ada+23] evaluated deep feature metrics for medial imaging and found that they correlate even better with radiologist's perceived evaluations.

It is important to note that not even radiologists agree in their assessment of image quality, illustrating the difficulty of evaluating the image quality of reconstructions.

As discussed, evaluating image quality in practice is difficult. In most current research papers, authors report the PSNR and SSIM to validate performance, and provide reconstruction examples to illustrate the performance of their algorithms. This is a reasonable practice, since the community is used to PSNR and SSIM, and together with illustrative example reconstructions, reporting those metrics enables to compare image reconstruction algorithms. Deep feature metrics are also a very good option to evaluate image reconstruction methods, but are not as simple to interpret, since their value depends on the particular choice of the neural network used to generate the feature representation.

14.4 Evaluating machine learning models

Machine learning models are typically evaluated as follows: We collect or generate a set of training examples, which for image reconstruction consist of target image $\mathbf{x}$ and measurement $\mathbf{y}$, $\{(\mathbf{x}_1, \mathbf{y}_1), ..., (\mathbf{x}_N, \mathbf{y}_N)\}$, shuffle the examples and split them into a test and training set. We then train models on the training set and evaluate the models on the test set, see Figure 14.6 for an illustration of this procedure. The hope is that evaluating the models on the test set gives a good estimate of the error when employing the model in practice.

If the test set is sufficiently large, and if we do not use the test set too often (for example by evaluating too many models on it), then the test error is a good estimate for the expected performance of the models on data that is distributed like the test data. If the test set is too small and if we evaluate a huge number of models on the test set, we might overfit to the test set. However, in practice, the test set is typically sufficiently large, and it is actually quite difficult to overfit to the test set.

However, the data that we apply a model to in practice and the data we test the model on are often not equally distributed. And a shift in distribution between two test sets (one reflecting the performance in practice, and one being our original test set) can result in a significant degradation of performance. A recent result demonstrating this is the following interesting experiment by Recht et al. [Rec+19] for image classifiers trained on the ImageNet dataset.

As mentioned before, ImageNet is a popular image classification task, where the goal is to classify an image as belonging to one of 1000 classes. Thousands of image classification methods have been evaluated on the ImageNet test set, which sparked fear of progress on this task being due to overfitting to the test set as opposed to real progress. This motivated Recht et al. [Rec+19] to collect an entirely new test set and to evaluate common algorithms on this new test set. Recht et al. [Rec+19] created a new test set by following the exact same procedure that was used to collect the original data as closely as possible, and evaluated a large number of deep neural networks on both the original ImageNet test set as well as on the new ImageNet test set. Recht et al. [Rec+19] found that while the accuracy of all methods dropped significantly on the new test set relative to the original one, the performance on both test sets is strongly correlated. Thus, a network that performs better than another network on the original test set also performed better on the new test set.

As a reason for the performance drop on the new relative to the original test set, Recht et al. [Rec+19] identified a distribution shift as the most likely cause. This finding illustrates that even if care is taken that a model is applied to data seemingly following a similar distribution, a distribution shift can occur in practice.

We next discuss distribution shifts in the context of image reconstruction problems. Darestani et al. [DCH21] recently studied the performance of several methods trained on the fastMRI dataset [Zbo+19] (a dataset consisting of knee measurements and reconstructions collected at New York University) but tested on a knee dataset from another university (Stanford), see Figure 14.6 for an illustration. Both datasets are from knees, but there are differences on how the data is obtained, for example the frequency resolution is slightly different. The paper evaluated three different classes of methods: Networks trained end-to-end, un-trained neural networks, and ℓ_1-minimization-based methods, and considered multiple variants of each methods (e.g., a variational network of different sizes, and a U-net of different sizes) in order to obtain a larger variety of models.

The results depicted in Figure 14.7 show a significant drop in performance when evaluated on a new set of images relative to evaluating on images from the same distribution as the training images, even though the new set of images is obtained only with slightly different scanner protocols in another hospital.

It can also be seen that the performance drop under the distribution shift is similar for all considered trained and un-trained methods. Thus, out-of-distribution performance is strongly correlated with in-distribution performance. This is surprising because un-trained methods depend only very mildly on the distribution through hyper-parameter tuning, and we might have expected them to be less sensitive to a distribution shift. As a consequence, the gain achieved by a given method over another typically retains this gain even under distribution shifts.

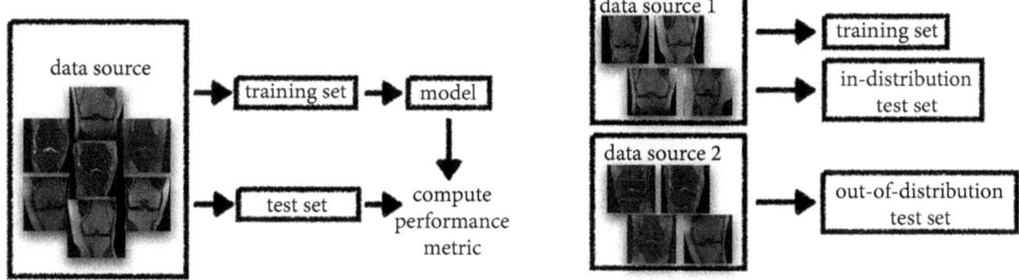

Figure 14.6 **Left:** In-distribution evaluation. Machine learning models are typically evaluated by constructing a non-overlapping training and test set from the same data source. **Right:** Out-of-distribution evaluation. Training and test set are from different data sources.

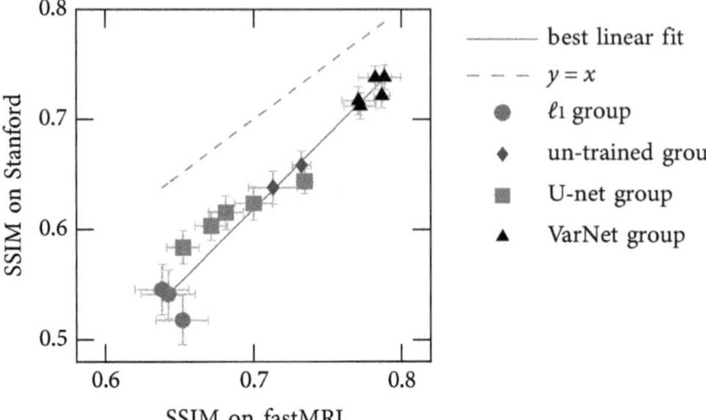

Figure 14.7 Reconstruction methods trained on one knee dataset (fastMRI) do not necessarily generalize well to another knee dataset (Stanford dataset), if there are differences in how the datasets are obtained. All of the considered reconstruction methods lose a similar amount of accuracy when evaluated on the Stanford set (through a dataset shift). The word "group" in the legend refers to variants of the respective method and error bars are 95% confidence intervals.

Distribution shifts like the one illustrated here can lead to significant degradation of performance in practice. There are two potential remedies. The first is to develop methods that adapt to a distribution shift, for example by tuning a regularization parameter for ℓ_1-regularized least-squares at inference. The second is to train a neural network on a diverse dataset, that way the test distribution might be close to the trained distribution and the performance loss due to a distribution shift is small.

14.5 Errors, hallucinations, and biases

Image reconstruction algorithms are based on image priors, either implicitly (e.g., networks trained in a supervised fashion) or explicitly (e.g., approaches based on GAN priors). Thus, image reconstruction algorithms interpolate to some extent based on those image priors. This can lead to hallucinations and errors. Errors and hallucinations can occur because there is too little information in the measurement, and thus the image prior is leveraged significantly to generate an image that is potentially far from the measurement and/or misleading. Errors and hallucinations can also occur when the original image to be reconstructed is far from the training or calibration data of the reconstruction algorithm.

In this section, we discuss a few examples of errors and hallucinations of deep learning-based image reconstruction methods, and end with a brief history of biases in photography, demonstrating that biases can also occur in traditional, entirely physics-based imaging technologies.

14.5.1 Errors and hallucinations

We start with an example of a super-resolution problem. We took a picture of President Biden and downscaled it significantly, see Figure 14.8, left panel. We then used the GAN-based PULSE [Men+20] algorithm to super-resolve the image. Recall that the PULSE algorithm imposes a GAN prior subject to the downscaled reconstructed image being consistent with the downsampled image. The GAN prior is trained on the CelebA-HQ dataset, a dataset that consists of 30k high-resolution faces of celebrities. We can see that the reconstructed person looks very different from Biden, and the person looks younger than Biden. This is not surprising: The downsampled image of Biden has very low resolution, but many of us can recognize Biden based on internal priors shaped by pictures of Biden we have seen before. The CelebA-HQ dataset, however, does not contain pictures of Biden and in addition the majority of training images is of younger individuals.

Next, we show an error occurring in medical image reconstruction. Figure 14.9 shows reconstructions of an image of a knee containing a pathology shown in the red square, along with reconstruction based on 4x and 8x undersampled measurement with ℓ_1-regularized least-squares as well as with a U-net trained on the fastMRI Knee dataset on 4x and 8x acceleration.

It can be seen that the reconstruction of the U-net looks good even for 8x acceleration. In terms of PSNR (or SSIM) and by visual inspection, the reconstructions of the U-net are better than the reconstructions with ℓ_1-regularized least-squares.

downsampled image

GAN-based reconstruction

Figure 14.8 A downsampled image of President Biden along with a GAN-based reconstruction illustrating biases and hallucination in image reconstruction.

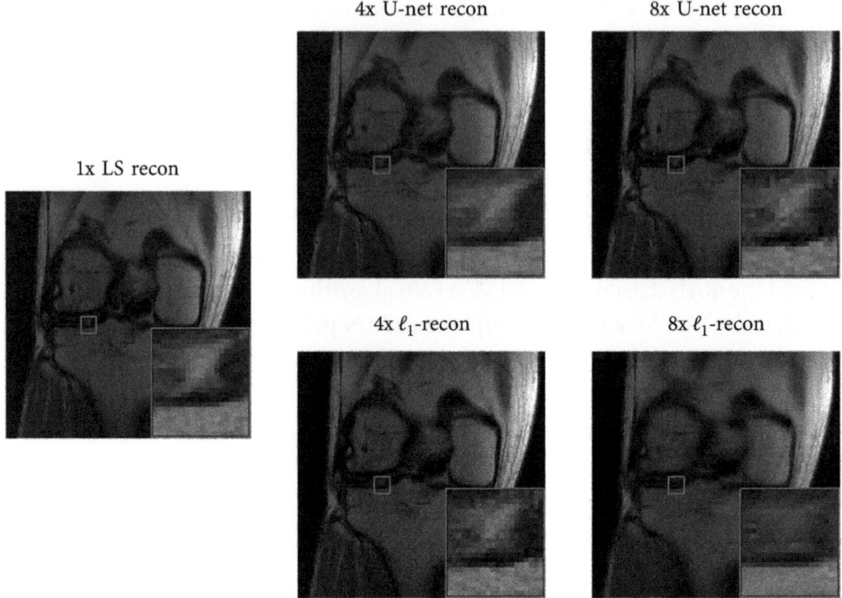

Figure 14.9 Reconstructions of a knee from the fastMRI+ dataset [Zha+22] with a pathology, specifically a cartilage damage. For 4x acceleration, the pathology is visible and for 8x acceleration it is not, neither for the ℓ_1-reconstruction nor for U-net reconstruction.

For both methods, at 8x acceleration, the pathology is not visible. However, the reconstruction of ℓ_1-regularized least-squares shows artifacts and is easier to identify as a bad reconstruction. The U-net reconstruction, however looks good and thus can be misleading.

This illustrates that better image priors that yield better image reconstructions, for example through training a U-net relative to using a sparsity prior through ℓ_1-regularization, can also lead to more realistic errors and can thus be misleading. To deal with this, it is important to use such methods only in a regime where they work well. This can be determined through testing and careful evaluation.

14.5.2 History of biases in photography

Biases can easily and sometimes unconsciously slip into imaging technologies, not just into AI-based imaging technologies. An example is colour calibration for colour-film photography and colour television. In the early days of colour photography and colour television, technicians used the image of a white woman with a colourful dress to calibrate colour so that the skin tone looks natural to a human eye. Such cards are called Shirley Cards, named after Shirley Page, one of the first

Figure 14.10 Kodak Shirley Card, picture from Rosa Menkman, licensed under CC BY 2.0.

models featured on the cards used by the company Kodak in the 1940s and 1950s to calibrate skin tones, colour, and exposure for their colour film processing. The history of those cards, how they can lead to biases, and how these cards developed throughout the years is detailed by Roth [Rot09].

Problems arose when taking pictures of groups of people with different skin tones with films calibrated based on early Shirley Cards (Figure 14.10). In such photographs, faces with whiter skins were represented better than faces with darker skins. Calibrating film chemistry, photo laboratory procedures, printing technologies, screen colours and digital cameras with the early Shirley Cards lead to a bias of those technologies being better able to capture white faces than coloured ones. Throughout the years, films have been redesigned in order to better represent different skin tones simultaneously, and the Shirley Cards have been redesigned to represent multiple ethnicities, for example Kodak introduced a multi-racial Shirley Card in 1995.

In principle, digital photography can somewhat lessen the problem of skin tone and other biases since each photo can be tailored individually. However, digital cameras have to explicitly account for being able to work for different skin tones. Current technologies for digital photography like Google's Real Tone are efforts to address skin tone biases and enable the camera to represent people of different skin tones accurately.

15
Advanced reconstruction problems

Throughout this book, we discussed deep learning-based reconstruction methods largely for 2D reconstruction problems in order to focus on the concepts of deep learning-based imaging, without the need to introduce too much complexity through the problem description. However, many important imaging problems are beyond the 2D imaging setups we largely focused on.

In this chapter, we will briefly discuss three instances of relatively advanced reconstruction problems: 3D and 4D (3D plus time) reconstruction and how some of the deep learning based techniques discussed in this book can be leveraged for those problems. The deep learning-based approaches discussed here are only examples of reconstruction approaches for those problems, a variety of methods for 3D and 4D reconstruction are actively developed.

15.1 3D reconstruction

In many medical, scientific, and other imaging systems the goal is to reconstruct a 3D volume from measurements. An example is cryogenic electron tomography discussed already in the introduction (Section 1.1.5). Another important example is 3D accelerated MRI.

For 3D accelerated MRI, the goal is to reconstruct a 3D volume $\mathbf{x} \in \mathbb{C}^{W \times H \times D}$ from undersampled measurements. The measurements are taken by several receiver coils $1, \ldots, C$, and are given by

$$\mathbf{y}_i = \mathbf{MFS}_i\mathbf{x} + \mathbf{Z}_i, \quad i = 1, \ldots, C, \tag{15.1}$$

where $\mathbf{S}_i$ is the sensitivity map associated with the i-th receiver coil, $\mathbf{F}$ is the 3D discrete Fourier transform, and $\mathbf{Z}_i$ is measurement noise. Moreover, $\mathbf{M}$ is an undersampling mask, which selects the frequencies that are measured by the scanner. The acceleration factor, i.e., the ratio of total frequencies over the frequencies that are measured, is typically higher for 3D scans than for 2D scans; for 2D scans acceleration factors of 2–4 are used, and for 3D scans acceleration factors are in the range 6–10.

As discussed earlier, for 2D accelerated MRI, neural networks trained in a supervised fashion, such as the unrolled variational network, perform best. We could, in principle, approach 3D accelerated MRI reconstruction by applying a

Deep Learning for Computational Imaging. Reinhard Heckel, Oxford University Press. © Reinhard Heckel (2025).
DOI: 10.1093/oso/9780198947172.003.0015

2D reconstruction method that reconstructs the x-y planes of the volume slice-by-slice. However, this is suboptimal as it does not leverage the structure in z-direction of the volume.

We could also, in principle, apply the supervised methods discussed here, for example train a 3D-version of an unrolled network. However, there is a lack of good 3D MRI target data for supervised learning, which is due to difficulties in collecting fully-sampled 3D MRI volumes required to generate ground-truth 3D target data. Collecting a fully-sampled MRI volume at a sufficiently high resolution to be able to generate a good target volume would take a very long time. Moreover, long scan times increase the likelihood of patient movements, which leads to motion artifacts. Scan times in MRI range from a few minutes to over an hour.

To circumvent this issue, and still be able to benefit from the strong priors deep networks can provide, one can leverage 2D priors. For example, it is possible to train a 2D-diffusion prior and leverage it for 3D reconstruction [Wu+24; Chu+23b]. This can be accomplished by building on the variational approach from Section 10.4.3 and optimizing a loss function of the form

$$\mathcal{L}(\mathbf{x}) = \sum_{i=1}^{C} \|\mathbf{y}_i - \mathbf{MFS}_i\mathbf{x}\|_2^2 + \mathbb{E}_{\mathbf{s}\sim\text{2D-slices}(\mathbf{x})}\left[R(\mathbf{s})\right], \qquad 15.2$$

where $R(\mathbf{s})$ is a residual denoiser training loss of a 2D-diffusion model (i.e., Equation (10.32)). Moreover, $\mathbf{s} \sim$ 2D-slices($\mathbf{x}$) means we choose 2D slices of the 3D volume $\mathbf{x}$ according to some distribution, e.g., uniformly at random from all x-y, x-z, and y-z slices. This approach works well for 3D reconstruction, and also illustrates how the techniques discussed in this book for a simple 2D reconstruction problem can be adapted to a more difficult setup.

15.2 Dynamic and 4D reconstruction

In many imaging problems of interest the object of interest is non-static.

Example 1: Cryo electron microscopy
Single particle cryo electron microscopy (cryo-EM) is a Nobel-prize winning method for determining the three-dimensional structure of proteins at near-atomic resolution.

In cryo-EM, a solution containing multiple copies of one protein (or more generally, a biological macromolecule) is frozen rapidly into a thin layer of ice. Proteins are three-dimensional structures, but their orientation in the layer of ice is unknown. The frozen sample is hit by an electron beam and a two-dimensional tomographic projection of the entire sample is recorded. From this projection of the 3D-sample, projected 2D-images of the different instances of the protein are extracted. Each of those projected 2D-images corresponds to a separate instance of the protein in an unknown orientation and is very noisy, which makes the problem of reconstructing a three-dimensional structure of the protein a challenging algorithmic problem. The projections are very noisy since the electron dose is low to prevent radiation damage of the proteins.

Many proteins exist in a non-static continuum of three-dimensional configurations (or 3D-shapes), which often correspond to different functions of the protein. This non-static nature or heterogeneity is a significant challenge for reconstruction, since each projected 3D-image can correspond to a different three-dimensional structure. Thus, the problem of reconstructing a protein that exists in a continuum of three-dimensional configurations is a dynamic inverse problem.

Zhong et al. [Zho+21] proposed a reconstruction approach based on variational autoencoders. The general idea is to represent the three-dimensional configuration of the protein with a distribution described by a variational autoencoder. Given are the 2D projections of the 3D protein in different unknown configurations along with estimates of the orientations of the proteins. The encoder maps the 2D projection to a latent variable, and a decoder takes the latent variable as well as the estimates of the orientations as input and maps them to a 3D volume representing the 3D protein. In the training step, the encoder and decoder are trained on the given particle images of a single probe along with the estimated image poses. After fitting, the decoder represents an estimate of the distribution of the 3D structure of the protein, including its various conformations. The variational autoencoder can then be leveraged to visualize and study the structure of the protein.

Example 2: Dynamic cardiac MRI

Another dynamic inverse problem is cine or dynamic cardiac MRI, an imaging technique for capturing a moving heart. Cine cardiac MRI enables the diagnosis of cardiac diseases and is non-invasive. The data acquisition process of MRI is inherently slow, and thus MRI traditionally has difficulties imaging moving organs, since during a scan the organ moves significantly. Indeed, motion during an MRI scan is a major source of image degradation, and even small movements result in strong artifacts [ZMH15].

In cardiac MRI, movement arises from cardiac motion (heart pumping), respiration (breathing), and patient movement. In current cardiac MRI protocols, patients are asked to periodically hold their breath. During those breath-hold periods, measurements are acquired throughout several cardiac cycles, and are synchronized via an electrocardiogram taken in parallel.

Many of the current reconstruction methods (both traditional and deep learning-based [Bus+20; Lei+19]) assume that the cardiac motion is periodic and often ignore respiratory motion, since patients hold their breath. Such binning-based reconstruction is illustrated in Figure 15.1, left panel: The MRI scanner collects measurements $\mathbf{y}_{c,k}$ during heart cycles $c = 1, \ldots, C$ and for time bins within

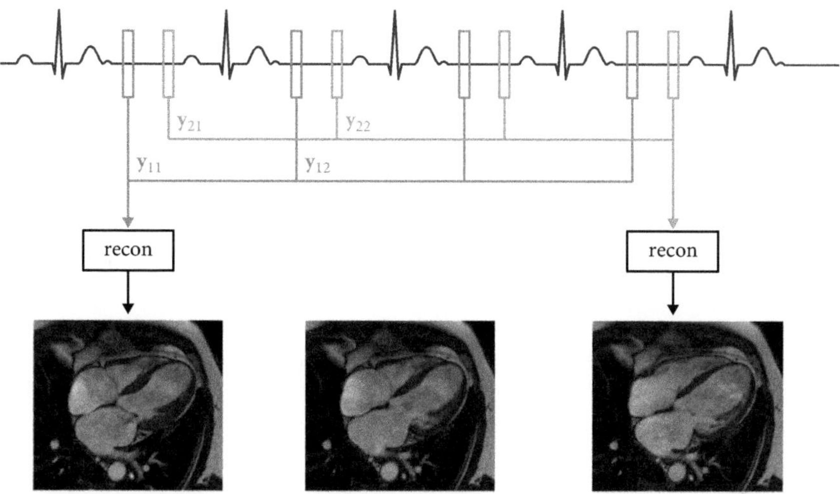

Figure 15.1 Traditional cine cardiac MRI protocols collect measurements during breath-hold periods over several cardiac cycles. Those measurements are assigned to a bin in the cardiac cycle and for each such bin, an image is reconstructed with a method that regards the image as static. This induces artifacts and limits resolution since the images are not static.

a heart cycle $k = 1, \ldots, K$. Using an electrocardiogram that is taken in parallel to the scan, the measurements corresponding to the same bin in the cardiac cycle are collected as $\mathbf{y}_k = [\mathbf{y}_{1,k}, \ldots, \mathbf{y}_{C,k}]$. Finally, a 2D or 3D reconstruction approach (deep learning-based or not) assumes the heart to be static within a bin and reconstructs an image for each bin. This yields a video of a beating heart.

However, this reconstruction approach assumes that the heart is at the same state during all cycles, which induces a systematic error that can limit the image resolution.

Deep learning can offer another solution. For example, Huang et al. and Kunz et al. [Hua+23; KRH24] proposed to use implicit neural networks to represent the image or volume representing the heart and fit an implicit neural network to the given data, as illustrated in Figure 15.2.

An instance of this approach is as follows: We simply take a neural representation, as introduced in Chapter 12, which takes a spatial (x, y) (or (x, y, z)) coordinate as well as a time coordinate t as input and maps this coordinate to a complex-valued pixel value. The coordinate-based neural representation consists of a Fourier-feature map (see Equation (12.1)) mapping the spatial coordinate to a feature representation, and another Fourier-feature map mapping the time coordinate to a feature representation. The feature representations of the coordinates are mapped to a complex-valued pixel value. See Figure 15.2, left panel, for an illustration of the network.

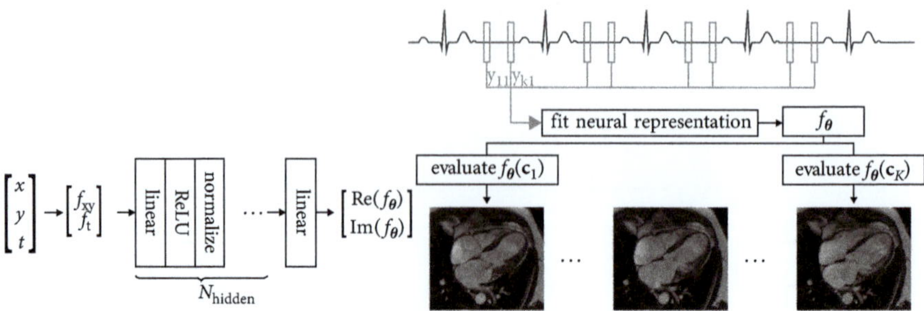

Figure 15.2 Reconstruction with a neural representation. We fit the neural representation shown on the left to be consistent with all the measurement data we have, which yields the representation f_θ of the heart. The representation can then be evaluated to reconstruct an image of the heart at different time points.

The network f_θ is fitted to the measurement data $\mathbf{y}_{11}, \mathbf{y}_{12}, \ldots, \mathbf{y}_{21}, \mathbf{y}_{22}, \ldots$, and after fitting we can evaluate the network to generate an image corresponding to a given time point of the heart's state. See Figure 15.2, right panel, for an illustration. This reconstruction approach regularizes through the architecture of the neural representation; the network has an implicit bias of nearby pixels and nearby points in time being close, as discussed in Chapter 12. The approach performs well compared to the traditional approach based on binning, and in addition does not require binning and thus does not require breath holding.

Yet, this approach is computationally relatively expensive, as fitting the neural representation is expensive. The approach leverages the implicit bias of neural representation well, but approaches that leverage-trained priors, for example generative priors, might perform significantly better.

15.3 Chapter notes

In this chapter we discussed three reconstruction problems beyond 2D image reconstruction, that illustrate the complexity of advanced imaging setups. While deep learning-based approaches for 2D imaging are already widely deployed today, for example in some medical imaging devices and consumer photography, for more advanced imaging problems like 3D and 4D reconstruction, as well as motion-corrected reconstruction, deep learning methods are still being actively developed.

APPENDIX

Mathematical background

In this section, we briefly review common notation and some concepts from calculus and probability that are frequently used in this book.

A.1 Notation

We start with defining common notation used throughout this book.

- Lower case letters x, η denote scalars.
- Lower case boldface letters $\mathbf{x}, \mathbf{y}, \mathbf{z}$ denote vectors and higher-dimensional signals, such as an image or a volume.
- Upper case boldface letters $\mathbf{A}, \mathbf{B}$ denote matrices typically acting as operators on vectors.
- The identity matrix is $\mathbf{I}$.
- Transpose. The transpose of a vector $\mathbf{b}$ or matrix $\mathbf{A}$ is denoted by $\mathbf{b}^T$ and $\mathbf{A}^T$.
- The inverse of a full-rank square matrix $\mathbf{A}$ is denoted by $\mathbf{A}^{-1}$.
- Pseudo-inverse. For a matrix with full column-rank, $\mathbf{A}^{\dagger} = (\mathbf{A}^T\mathbf{A})^{-1}\mathbf{A}^T$ is the pseudo-inverse of the matrix.
- Inner product: $<\mathbf{x}, \mathbf{y} = \mathbf{x}^{\mathrm{T}}\mathbf{y}>$ denotes the inner product of the vectors $\mathbf{x}$ and $\mathbf{y}$.

A.2 Linear algebra

Gradient

Consider a multi-variate function $f: \mathbb{R}^d \to \mathbb{R}$. The gradient of the function f at a point $\boldsymbol{\theta}$ is the vector of partial derivatives

$$\nabla f(\boldsymbol{\theta}) = \begin{bmatrix} \frac{\partial f(\theta)}{\partial \theta_1} \\ \frac{\partial f(\theta)}{\partial \theta_2} \\ \cdots \\ \frac{\partial f(\theta)}{\partial \theta_d} \end{bmatrix}.$$

Sometimes we consider a function depending on multiple variables and write $\nabla_{\boldsymbol{\theta}} f(\boldsymbol{\theta})$ to be explicit about which functional argument we refer to.

Jacobian

Consider a multi-variate mapping $\mathbf{f}: \mathbb{R}^d \to \mathbb{R}^n$. The matrix $\mathbf{J}_{\boldsymbol{\theta}} \in \mathbb{R}^{n\times d}$ which contains the partial derivatives of the multi-variate mapping at $\boldsymbol{\theta}$ is called the Jacobian matrix:

$$\mathbf{J}_{\boldsymbol{\theta}}\mathbf{f}(\boldsymbol{\theta}) = \begin{bmatrix} \frac{\partial f_1(\theta)}{\partial \theta_1} & \cdots & \frac{\partial f_1(\theta)}{\partial \theta_d} \\ \vdots & \ddots & \vdots \\ \frac{\partial f_n(\theta)}{\partial \theta_1} & \cdots & \frac{\partial f_n(\theta)}{\partial \theta_d} \end{bmatrix}.$$

Singular value decomposition (SVD)

The singular value decomposition is a fundamental matrix factorization technique. It decomposes a matrix $\mathbf{A} \in \mathbb{R}^{m \times n}$ into the product of an orthonormal matrix $\mathbf{U} \in \mathbb{R}^{m \times m}$, a diagonal matrix $\mathbf{\Sigma} \in \mathbb{R}^{m \times n}$, and another orthonormal matrix $\mathbf{V} \in \mathbb{R}^{n \times n}$ as

$$\mathbf{A} = \mathbf{U}\mathbf{\Sigma}\mathbf{V}^T.$$

The matrix $\mathbf{U}$ contains the left-singular vector of the matrix $\mathbf{A}$, the matrix $\mathbf{\Sigma}$ contains the singular values on its diagonal, and $\mathbf{V}$ contains the right-singular vectors as columns.

The singular values are the square roots of the eigenvalues of the matrix $\mathbf{A}^T\mathbf{A}$ or $\mathbf{A}\mathbf{A}^T$, since $\mathbf{A}^T\mathbf{A} = \mathbf{V}\mathbf{\Sigma}^T\mathbf{\Sigma}\mathbf{V}^T$ and $\mathbf{A}\mathbf{A}^T = \mathbf{U}\mathbf{\Sigma}\mathbf{\Sigma}^T\mathbf{U}^T$.

The SVD can be interpreted as decomposing a linear transformation represented by the matrix $\mathbf{A}$ into a rotation implemented by $\mathbf{V}^T$, a scaling implemented by $\mathbf{\Sigma}$, and another rotation implemented by $\mathbf{U}$.

Positive definite and positive semidefinite matrices

A positive definite matrix $\mathbf{A}$ is a square matrix such that for any non-zero vector $\mathbf{x}$, the scalar $\mathbf{x}^T\mathbf{A}\mathbf{x}$ is positive. This is the case if and only if all eigenvalues of $\mathbf{A}$ are positive.

A positive semidefinite matrix $\mathbf{A}$ is a square matrix such that for any non-zero vector $\mathbf{x}$, the scalar $\mathbf{x}^T\mathbf{A}\mathbf{x}$ is non-negative, which is equal to saying that all eigenvalues are non-negative.

Matrix and vector norms

The ℓ_2-norm of a vector $\mathbf{x}$ is defined as $\|\mathbf{x}\|_2 = \sqrt{\sum_i x_i^2}$, where x_i are the entries of $\mathbf{x}$. The ℓ_1-norm is defined as $\|\mathbf{x}\|_1 = \sum_i |x_i|$.

For a matrix $\mathbf{A}$, the standard operator norm, also known as the spectral norm, is defined as

$$\|\mathbf{A}\| = \sup_{\mathbf{x} \neq 0} \frac{\|\mathbf{A}\mathbf{x}\|_2}{\|\mathbf{x}\|_2}.$$

The norm can also be characterized in terms of the singular values of the matrix $\mathbf{A}$, specifically it is equal to the largest singular value, i.e., $\|\mathbf{A}\| = \max_i \sigma_i$. To see this, note that

$$\|\mathbf{A}\| = \sup_{\|\mathbf{x}\|_2=1} \|\mathbf{A}\mathbf{x}\|_2 = \sup_{\|\mathbf{x}\|_2=1} \|\mathbf{U}\mathbf{\Sigma}\mathbf{V}^T\mathbf{x}\|_2 = \sup_{\|\mathbf{x}'\|_2=1} \|\mathbf{\Sigma}\mathbf{x}'\|_2 = \max_i \sigma_i. \tag{16.1}$$

Here, we used the SVD and that the matrices $\mathbf{U}$ and $\mathbf{V}$ are orthonormal. The last inequality follows since the vector $\mathbf{x}'$ of norm one that maximizes $\|\mathbf{\Sigma}\mathbf{x}'\|_2$ is one at the index of the maximal singular value, and zero everywhere else.

A.3 Probability

Machine learning and inverse problem use probability to quantify uncertainty and to model noise. We review some basics we use throughout this book. The basic concept in probability is the notion of randomness. The outcome of certain experiments is not predictable. For example, in the context of imaging, the exact number of photons that hit a pixel on a sensor at a given time is random.

In such an experiment, there is a space Ω of possible outcomes, and each individual outcome is described by a point $\omega \in \Omega$. Subsets of the space of possible outcomes, Ω, are called events. For example, the number of photons arriving on a sensor during some time interval being larger than 100 is an event.

Random variables and probability distributions

Random variables characterize the outcomes of a random experiment. There are two main types of random variables: Discrete random variables take on a countable number of distinct values, an example is the number of photons that arrive at a sensor in a given time interval, and another a classical example is the outcome of a coin toss. Continuous random variables take on an uncountable number of values. An example is random measurement noise. A random vector is a random variable with sample space $\Omega = \mathbb{R}^d$.

A probability distribution describes how the values of a random variable are distributed, and provides a complete description of the probabilities association with all possible outcomes of a random variable.

We primarily use continuous random variables in this book. The probability distribution of a continuous random variable can be characterized by a probability density function p. A probability density function has the following three properties. It is non-negative, i.e., $p(x) \geq 0$ for all x, it is normalized to one, i.e., $\int_\Omega p(x)dx = 1$, and the probability that the random variable x falls within an interval $[a, b]$ is the integral over the pdf of that interval, i.e.,

$$\mathrm{P}[x \in [a, b]] = \int_a^b p(x)dx.$$

Expectation

The expectation of a continuous random variable x is defined as

$$\mathbb{E}[x] = \int_\Omega xp(x)dx.$$

A function of a random variable, $f(x)$ is also a random variable, and the expectation of the random variable $f(x)$ is defined as

$$\mathbb{E}[f(x)] = \int_\Omega f(x)p(x)dx.$$

Important examples are the mean of a random vector

$$\boldsymbol{\mu} = \mathbb{E}[\mathbf{x}],$$

and the co-variance matrix of a random vector

$$\boldsymbol{\Sigma} = \mathbb{E}\left[(\mathbf{x} - \boldsymbol{\mu})(\mathbf{x} - \boldsymbol{\mu})\right]^T.$$

The expectation of a random variable may not exist (such as for the Cauchy distribution), but for the random variables considered in this book, expectations exist.

Gaussian random variables

An important example of a probability distribution used throughout this book is a Gaussian probability distribution $\mathcal{N}(\boldsymbol{\mu}, \boldsymbol{\Sigma})$ parameterized by the mean $\boldsymbol{\mu} \in \mathbb{R}^d$ and the co-variance matrix $\boldsymbol{\Sigma} \in \mathbb{R}^d$. A Gaussian random vector $\mathbf{x} \in \mathbb{R}^d$ has probability density

$$p(\mathbf{x}) = \frac{1}{\det\sqrt{2\pi\boldsymbol{\Sigma}}} \exp\left(-\frac{1}{2}(\mathbf{x} - \boldsymbol{\mu})^T \boldsymbol{\Sigma}^{-1}(\mathbf{x} - \boldsymbol{\mu})\right).$$

Gaussian random variables are also often called normal random variables, and we talk about a standard normal random variable if the mean is zero and the co-variance matrix is identity, i.e., $\mathcal{N}(0, \mathbf{I})$.

Gaussian random variables are so important for inverse problem and beyond because the average of many independent random variables looks like a Gaussian random variable.

Conditional probability and Bayes' Rule

The conditional probability is defined as the probability of an event A occurring given that another event B has already occurred. The conditional probability of event A given event B is defined as

$$\mathrm{P}[A|B] = \frac{\mathrm{P}[A \cap B]}{\mathrm{P}[B]},$$

where $\mathrm{P}[A \cap B]$ is the probability that both events A and B occur. Note that this definition assumes that the probability of event B occurring is non-zero, and the conditional probability is not defined if the probability of event B occurring is zero, i.e., when $\mathrm{P}[B] = 0$.

Bayes' Rule, used frequently in this book, follows directly from the definition of conditional probability and states that

$$\mathrm{P}[A|B] = \frac{\mathrm{P}[B \cap A]\mathrm{P}[A]}{\mathrm{P}[B]}.$$

For probability distribution, a conditional distribution and Bayes' theorem are analogous. The conditional density function of x given y is defined as

$$p(x|y) = \frac{p(x, y)}{p(y)},$$

where $p(x, y)$ is the joint density function of x and y, and $p(y)$ is the density function of the random variable y. Bayes' theorem states that

$$p(x|y) = \frac{p(y|x)p(x)}{p(y)}.$$

References

[ALB18] A. Abdelhamed, S. Lin, and M. S. Brown. "A High-Quality Denoising Dataset for Smartphone Cameras". In: *IEEE Conference on Computer Vision and Pattern Recognition*. 2018, pp. 1692–1700.

[Ada+23] P. M. Adamson et al. "Using Deep Feature Distances for Evaluating MR Image Reconstruction Quality". In: *NeurIPS 2023 Workshop on Deep Learning and Inverse Problems*. 2023.

[AT17] E. Agustsson and R. Timofte. "NTIRE 2017 Challenge on Single Image Super-Resolution: Dataset and Study". In: *IEEE Conference on Computer Vision and Pattern Recognition Workshops (CVPRW)*. 2017, pp. 1122–1131.

[AEB06] M. Aharon, M. Elad, and A. Bruckstein. "K-SVD: An Algorithm for Designing Overcomplete Dictionaries for Sparse Representation". In: *IEEE Transactions on Signal Processing* 54.11 (2006), pp. 4311–4322.

[Alo03] N. Alon. "Problems and Results in Extremal Combinatorics—I". In: *Discrete Mathematics* 273.1–3 (2003), pp. 31–53.

[AB18] J. Anaya and A. Barbu. "RENOIR – A Dataset for Real Low-Light Image Noise Reduction". In: *Journal of Visual Communication and Image Representation* 51 (2018), pp. 144–154.

[And82] B. D. O. Anderson. "Reverse-Time Diffusion Equation Models". In: *Stochastic Processes and their Applications* 12.3 (1982), pp. 313–326.

[Ant+18] N. Antipa, G. Kuo, R. Heckel, B. Mildenhall, E. Bostan, R. Ng, and L. Waller. "DiffuserCam: Lensless Single-Exposure 3D Imaging". In: *Optica* 5.1 (2018), pp. 1–9.

[Ant+20] V. Antun, F. Renna, C. Poon, B. Adcock, and A. C. Hansen. "On Instabilities of Deep Learning in Image Reconstruction and the Potential Costs of AI". In: *Proceedings of the National Academy of Sciences* (2020).

[ACB17] M. Arjovsky, S. Chintala, and L. Bottou. "Wasserstein Generative Adversarial Networks". In: *International Conference on Machine Learning*. 2017, pp. 214–223.

[Asi+20] M. Asim, M. Daniels, O. Leong, A. Ahmed, and P. Hand. "Invertible Generative Models for Inverse Problems: Mitigating Representation Error and Dataset Bias". In: *International Conference on Machine Learning*. 2020.

[BKH16] J. L. Ba, J. R. Kiros, and G. E. Hinton. "Layer Normalization". In: *arXiv:1607.06450 [cs, stat]* (2016).

[Bar+08] R. Baraniuk, M. Davenport, R. DeVore, and M. Wakin. "A Simple Proof of the Restricted Isometry Property for Random Matrices". In: *Constructive Approximation* 28.3 (2008), pp. 253–263.

[BJ03] R. Basri and D. Jacobs. "Lambertian Reflectance and Linear Subspaces". In: *IEEE Transactions on Pattern Analysis and Machine Intelligence* 25.2 (2003), pp. 218–233.

[BR19] J. Batson and L. Royer. "Noise2Self: Blind Denoising by Self-Supervision". In: *International Conference on Machine Learning*. 2019.

[Bay+18] A. G. Baydin, B. A. Pearlmutter, A. A. Radul, and J. M. Siskind. "Automatic Differentiation in Machine Learning: A Survey". In: *Journal of Machine Learning Research* 18.153 (2018), pp. 1–43.

[Bor+17] A. Bora, A. Jalal, E. Price, and A. G. Dimakis. "Compressed Sensing Using Generative Models". In: *International Conference on Machine Learning*. 2017.

[Bos+20] E. Bostan, R. Heckel, M. Chen, M. Kellman, and L. Waller. "Deep Phase Decoder: Self-calibrating Phase Microscopy with an Untrained Deep Neural Network". In: *Optica* (2020).

[BV04] S. Boyd and L. Vandenberghe. *Convex Optimization*. Cambridge University Press, 2004.

[Boy+11] S. Boyd, N. Parikh, E. Chu, B. Peleato, and J. Eckstein. "Distributed Optimization and Statistical Learning via the Alternating Direction Method of Multipliers". In: *Foundations and Trends in Machine Learning* 3.1 (2011), pp. 1–122.

[Bro+19] T. Brooks, B. Mildenhall, T. Xue, J. Chen, D. Sharlet, and J. T. Barron. "Unprocessing Images for Learned Raw Denoising". In: *IEEE Conference on Computer Vision and Pattern Recognition*. 2019.

[BSH12] H. C. Burger, C. J. Schuler, and S. Harmeling. "Image Denoising: Can Plain Neural Networks Compete with BM3D?" In: *IEEE Conference on Computer Vision and Pattern Recognition*. 2012.

[Bus+20] A. Bustin, N. Fuin, R. M. Botnar, and C. Prieto. "From Compressed-Sensing to Artificial Intelligence-Based Cardiac MRI Reconstruction". In: *Frontiers in Cardiovascular Medicine* 7 (2020).

[CT06] E. J Candès and T. Tao. "Near-Optimal Signal Recovery From Random Projections: Universal Encoding Strategies?" In: *IEEE Transactions on Information Theory* 52.12 (2006), pp. 5406–5425.

[CRT06a] E. J. Candès, J. K. Romberg, and T. Tao. "Robust Uncertainty Principles: Exact Signal Reconstruction from Highly Incomplete Frequency Information". In: *IEEE Transactions on Information Theory* 52.2 (2006), pp. 489–509.

[CP11] E. J. Candès and Y. Plan. "A Probabilistic and RIPless Theory of Compressed Sensing". In: *IEEE Transactions on Information Theory* 57.11 (2011).

[CRT06b] E. J. Candès, J. K. Romberg, and T. Tao. "Stable Signal Recovery from Incomplete and Inaccurate Measurements". In: *Communications on Pure and Applied Mathematics* 59.8 (2006), pp. 1207–1223.

[Cha+12] V. Chandrasekaran, B. Recht, P. A. Parrilo, and A. S. Willsky. "The Convex Geometry of Linear Inverse Problems". In: *Foundations of Computational Mathematics* 12.6 (2012), pp. 805–849.

[Che+21] H. Chen, Y. Wang, T. Guo, C. Xu, Y. Deng, Z. Liu, S. Ma, C. Xu, C. Xu, and W. Gao. "Pre-Trained Image Processing Transformer". In: *IEEE Conference on Computer Vision and Pattern Recognition*. 2021.

[CDS+98] S. S. Chen, D. L. Donoho, and M. A. Saunders. "Atomic Decomposition by Basis Pursuit". In: *SIAM Journal on Scientific Computing* 20.1 (1998), pp. 33–61.

[CYP15] Y. Chen, W. Yu, and T. Pock. "On Learning Optimized Reaction Diffusion Processes for Effective Image Restoration". In: *IEEE Conference on Computer Vision and Pattern Recognition*. 2015.

[Chu+23a] H. Chung, J. Kim, M. T. Mccann, M. L. Klasky, and J. C. Ye. "Diffusion Posterior Sampling for General Noisy Inverse Problems". In: *International Conference for Learning Representations*. 2023.

[Chu+23b] H. Chung, D. Ryu, M. T. McCann, M. L. Klasky, and J. C. Ye. "Solving 3D Inverse Problems Using Pre-Trained 2D Diffusion Models". In: *IEEE Conference on Computer Vision and Pattern Recognition*. 2023.

[CLH18] J. P. Cohen, M. Luck, and S. Honari. "Distribution Matching Losses Can Hallucinate Features in Medical Image Translation". In: *Medical Image Computing & Computer Assisted Intervention*. 2018.

[CK19] D. Colton and R. Kress. *Inverse Acoustic and Electromagnetic Scattering Theory*. Vol. 93. Applied Mathematical Sciences. Cham: Springer International Publishing, 2019.

[Dab+07] K. Dabov, A. Foi, V. Katkovnik, and K. Egiazarian. "Image Denoising by Sparse 3-D Transform-Domain Collaborative Filtering". In: *IEEE Transactions on Image Processing* 16.8 (2007), pp. 2080–2095.

[DH21] M. Z. Darestani and R. Heckel. "Accelerated MRI With Un-Trained Neural Networks". In: *IEEE Transactions on Computational Imaging* 7 (2021), pp. 724–733.

[DCH21] M. Z. Darestani, A. Chaudhari, and R. Heckel. "Measuring Robustness in Deep Learning Based Compressive Sensing". In: *International Conference on Machine Learning*. 2021.

[DG03] S. Dasgupta and A. Gupta. "An Elementary Proof of a Theorem of Johnson and Lindenstrauss". In: *Random Structures & Algorithms* 22.1 (2003), pp. 60–65.

[Den+09] J. Deng, W. Dong, R. Socher, L.-J. Li, K. Li, and L. Fei-Fei. "Imagenet: A Large-Scale Hierarchical Image Database". In: *IEEE Conference on Computer Vision and Pattern Recognition*. 2009.

[DSB17] L. Dinh, J. Sohl-Dickstein, and S. Bengio. "Density Estimation Using Real NVP". In: *International Conference for Learning Representations*. 2017.

[DKB15] L. Dinh, D. Krueger, and Y. Bengio. "NICE: Non-linear Independent Components Estimation". In: *arXiv:1410.8516 [cs]* (2015).

[Don+16] C. Dong, C. C. Loy, K. He, and X. Tang. "Image Super-Resolution Using Deep Convolutional Networks". In: *IEEE Transactions on Pattern Analysis and Machine Intelligence* 38.2 (2016), pp. 295–307.

[Dos+21] A. Dosovitskiy et al. "An Image Is Worth 16x16 Words: Transformers for Image Recognition at Scale". In: *International Conference for Learning Representations*. 2021.

[Du+18] S. S. Du, X. Zhai, B. Poczos, and A. Singh. "Gradient Descent Provably Optimizes Over-parameterized Neural Networks". In: *International Conference on Learning Representations*. 2018.

[DHS11] J. Duchi, E. Hazan, and Y. Singer. "Adaptive Subgradient Methods for Online Learning and Stochastic Optimization". In: *Journal of Machine Learning Research* 12.Jul (2011), pp. 2121–2159.

[Dup+21] E. Dupont, A. Goliński, M. Alizadeh, Y. W. Teh, and A. Doucet. "COIN: COmpression with Implicit Neural Representations". In: *arXiv:2103.03123 [cs, eess]* (2021).

[EA06] M. Elad and M. Aharon. "Image Denoising Via Sparse and Redundant Representations Over Learned Dictionaries". In: *IEEE Transactions on Image Processing* 15.12 (2006), pp. 3736–3745.

[Eld09] Y. C. Eldar. "Generalized SURE for Exponential Families: Applications to Regularization". In: *IEEE Transactions on Signal Processing* 57.2 (2009), pp. 471–481.

[FTS22] Z. Fabian, B. Tinaz, and M. Soltanolkotabi. "HUMUS-Net: Hybrid Unrolled Multiscale Network Architecture for Accelerated MRI Reconstruction". In: *Advances in Neural Information Processing Systems* (2022).

[FMF16] A. Fawzi, S.-M. Moosavi-Dezfooli, and P. Frossard. "Robustness of Classifiers: From Adversarial to Random Noise". In: *Advances in Neural Information Processing Systems*. 2016.

[FR13] S. Foucart and Rauhut, Holger. *A Mathematical Introduction to Compressive Sensing*. Springer Berlin Heidelberg, 2013.

[Gen+22] M. Genzel, I. Gühring, J. Macdonald, and M. März. "Near—-Exact Recovery for Tomographic Inverse Problems via Deep Learning". In: *International Conference on Machine Learning*. 2022.

[GMM22] M. Genzel, J. Macdonald, and M. Marz. "Solving Inverse Problems With Deep Neural Networks – Robustness Included". In: *IEEE Transactions on Pattern Analysis and Machine Intelligence* (2022).

[Goo+14] I. Goodfellow, J. Pouget-Abadie, M. Mirza, B. Xu, D. Warde-Farley, S. Ozair, A. Courville, and Y. Bengio. "Generative Adversarial Nets". In: *Advances in Neural Information Processing Systems 27*. 2014.

[GL10] K. Gregor and Y. LeCun. "Learning Fast Approximations of Sparse Coding". In: *International Conference on Machine Learning*. 2010.

[Ham+18] K. Hammernik, T. Klatzer, E. Kobler, M. P. Recht, D. K. Sodickson, T. Pock, and F. Knoll. "Learning a Variational Network for Reconstruction of Accelerated MRI Data". In: *Magnetic Resonance in Medicine* 79.6 (2018), pp. 3055–3071.

[HH19] R. Heckel and P. Hand. "Deep Decoder: Concise Image Representations from Untrained Non-convolutional Networks". In: *International Conference on Learning Representations.* 2019.

[HS20a] R. Heckel and M. Soltanolkotabi. "Compressive Sensing with Un-Trained Neural Networks: Gradient Descent Finds the Smoothest Approximation". In: *International Conference on Machine Learning.* 2020.

[HS20b] R. Heckel and M. Soltanolkotabi. "Denoising and Regularization via Exploiting the Structural Bias of Convolutional Generators". In: *International Conference on Learning Representations.* 2020.

[HJA20] J. Ho, A. Jain, and P. Abbeel. "Denoising Diffusion Probabilistic Models". In: *Neural Information Processing Systems.* 2020.

[Hua+21] T. Huang, S. Li, X. Jia, H. Lu, and J. Liu. "Neighbor2Neighbor: Self-Supervised Denoising from Single Noisy Images". In: *IEEE Conference on Computer Vision and Pattern Recognition.* 2021.

[Hua+23] W. Huang, H. Li, G. Cruz, J. Pan, D. Rueckert, and K. Hammernik. "Neural Implicit K-Space for Binning-free Non-Cartesian Cardiac MR Imaging". In: *Information Processing in Medical Imaging.* 2023.

[Hua+18] Y. Huang, T. Würfl, K. Breininger, L. Liu, G. Lauritsch, and A. Maier. "Some Investigations on Robustness of Deep Learning in Limited Angle Tomography". In: *Medical Image Computing and Computer Assisted Intervention.* 2018.

[HA20] R. Hyder and M. S. Asif. "Generative Models for Low-Dimensional Video Representation and Reconstruction". In: *IEEE Transactions on Signal Processing* 68 (2020), pp. 1688–1701.

[IS15] S. Ioffe and C. Szegedy. "Batch Normalization: Accelerating Deep Network Training by Reducing Internal Covariate Shift". In: *International Conference on Machine Learning.* 2015.

[JGH18] A. Jacot, F. Gabriel, and C. Hongler. "Neural Tangent Kernel: Convergence and Generalization in Neural Networks". In: *Neural Information Processing Systems.* 2018.

[JH19] G. Jagatap and C. Hegde. "Algorithmic Guarantees for Inverse Imaging with Untrained Network Priors". In: *Neural Information Processing Systems.* 2019.

[JS08] V. Jain and S. Seung. "Natural Image Denoising with Convolutional Networks". In: *Neural Information Processing Systems.* 2008.

[Jal+21] A. Jalal, M. Arvinte, G. Daras, E. Price, A. G. Dimakis, and J. I. Tamir. "Robust Compressed Sensing MRI with Deep Generative Priors". In: *Neural Information Processing Systems.* 2021.

[Jin+17] K. H. Jin, M. T. McCann, E. Froustey, and M. Unser. "Deep Convolutional Neural Network for Inverse Problems in Imaging". In: *IEEE Transactions on Image Processing* 26.9 (2017), pp. 4509–4522.

[KS05] J. Kaipio and E. Somersalo. *Statistical and Computational Inverse Problems.* Vol. 160. Applied Mathematical Sciences. New York: Springer-Verlag, 2005.

[KMW17] U. S. Kamilov, H. Mansour, and B. Wohlberg. "A Plug-and-Play Priors Approach for Solving Nonlinear Imaging Inverse Problems". In: *IEEE Signal Processing Letters* 24.12 (2017), pp. 1872–1876.

[Kam+23] U. S. Kamilov, C. A. Bouman, G. T. Buzzard, and B. Wohlberg. "Plug-and-Play Methods for Integrating Physical and Learned Models in Computational Imaging". In: *IEEE Signal Processing Magazine* 40.1 (2023), pp. 85–97.

[Ker+23] B. Kerbl, G. Kopanas, T. Leimkuehler, and G. Drettakis. "3D Gaussian Splatting for Real-Time Radiance Field Rendering". In: *ACM Transactions on Graphics* 42.4 (2023), pp. 1–14.

[KB15] D. P. Kingma and J. Ba. "Adam: A Method for Stochastic Optimization". In: *International Conference for Learning Representations.* 2015.

[KD18] D. P. Kingma and P. Dhariwal. "Glow: Generative Flow with Invertible 1x1 Convolutions". In: *Neural Information Processing Systems.* 2018.

[KW19] D. P. Kingma and M. Welling. "An Introduction to Variational Autoencoders". In: *Foundations and Trends in Machine Learning* 12.4 (2019), pp. 307–392.

[KW14] D. P. Kingma and M. Welling. "Auto-Encoding Variational Bayes". In: *International Conference on Learning Representations.* 2014.

[Kin+21] D. P. Kingma, T. Salimans, B. Poole, and J. Ho. "Variational Diffusion Models". In: *Advances in Neural Information Processing Systems.* 2021.

[KH23] T. Klug and R. Heckel. "Scaling Laws For Deep Learning Based Image Reconstruction". In: *International Conference for Learning Representations.* 2023.

[Kno+20] F. Knoll et al. "Advancing Machine Learning for MR Image Reconstruction with an Open Competition: Overview of the 2019 fastMRI Challenge". In: *Magnetic Resonance in Medicine* 84.6 (2020), pp. 3054–3070.

[KW11] F. Krahmer and R. Ward. "New and Improved Johnson-Lindenstrauss Embeddings via the Restricted Isometry Property". In: *SIAM Journal on Mathematical Analysis* 43.3 (2011), pp. 1269–1281.

[KBJ19] A. Krull, T.-O. Buchholz, and F. Jug. "Noise2Void - Learning Denoising From Single Noisy Images". In: *IEEE Conference on Computer Vision and Pattern Recognition.* 2019.

[KRH24] J. F. Kunz, S. Ruschke, and R. Heckel. "Implicit Neural Networks with Fourier-Feature Inputs for Free-breathing Cardiac MRI Reconstruction". In: *IEEE Transactions on Computational Imaging* (2024).

[Leh+18] J. Lehtinen, J. Munkberg, J. Hasselgren, S. Laine, T. Karras, M. Aittala, and T. Aila. "Noise2Noise: Learning Image Restoration without Clean Data". In: *International Conference on Machine Learning.* 2018.

[Lei+19] T. Leiner, D. Rueckert, A. Suinesiaputra, B. Baeßler, R. Nezafat, I. Išgum, and A. A. Young. "Machine Learning in Cardiovascular Magnetic Resonance: Basic Concepts and Applications". In: *Journal of Cardiovascular Magnetic Resonance* 21.1 (2019), p. 61.

[Lia+21] J. Liang, J. Cao, G. Sun, K. Zhang, L. Van Gool, and R. Timofte. "SwinIR: Image Restoration Using Swin Transformer". In: *International Conference on Computer Vision.* 2021.

[LH21] K. Lin and R. Heckel. "Vision Transformers Enable Fast and Robust Accelerated MRI". In: *Medical Imaging with Deep Learning.* 2021.

[LDP07] M. Lustig, D. Donoho, and J. M. Pauly. "Sparse MRI: The Application of Compressed Sensing for Rapid MR Imaging". In: *Magnetic Resonance in Medicine* 58.6 (2007), pp. 1182–1195.

[Mal08] S. Mallat. *A Wavelet Tour of Signal Processing: The Sparse Way.* Academic Press, 2008.

[MLH22] Y. Mansour, K. Lin, and R. Heckel. *Image-to-Image MLP-mixer for Image Reconstruction.* 2022. arXiv: 2202.02018.

[Mar+23] M. Mardani, J. Song, J. Kautz, and A. Vahdat. "A Variational Perspective on Solving Inverse Problems with Diffusion Models". In: *International Conference for Learning Representations.* 2023.

[Mar+01] D. Martin, C. Fowlkes, D. Tal, and J. Malik. "A Database of Human Segmented Natural Images and Its Application to Evaluating Segmentation Algorithms and Measuring Ecological Statistics". In: *IEEE International Conference on Computer Vision.* 2001.

[Mas+20] A. Mason, J. Rioux, S. E. Clarke, A. Costa, M. Schmidt, V. Keough, T. Huynh, and S. Beyea. "Comparison of Objective Image Quality Metrics to Expert Radiologists' Scoring of Diagnostic Quality of MR Images". In: *IEEE Transactions on Medical Imaging* 39.4 (2020), pp. 1064–1072.

[MPT08] S. Mendelson, A. Pajor, and N. Tomczak-Jaegermann. "Uniform Uncertainty Principle for Bernoulli and Subgaussian Ensembles". In: *Constructive Approximation* 28.3 (2008), pp. 277–289.

[Men+20] S. Menon, A. Damian, S. Hu, N. Ravi, and C. Rudin. "PULSE: Self-Supervised Photo Upsampling via Latent Space Exploration of Generative Models". In: *Conference on Computer Vision and Pattern Recognition.* 2020.

[Met+20] C. A. Metzler, A. Mousavi, R. Heckel, and R. G. Baraniuk. "Unsupervised Learning with Stein's Unbiased Risk Estimator". In: *International Biomedical and Astronomical Signal Processing (BASP) Frontiers Workshop*. 2020.

[Mil+20] B. Mildenhall, P. P. Srinivasan, M. Tancik, J. T. Barron, R. Ramamoorthi, and R. Ng. "NeRF: Representing Scenes as Neural Radiance Fields for View Synthesis". In: *European Conference on Computer Vision*. 2020.

[Muc+21] M. J. Muckley et al. "Results of the 2020 fastMRI Challenge for Machine Learning MR Image Reconstruction". In: *IEEE Transactions on Medical Imaging* 40.9 (2021), pp. 2306–2317.

[Mül+22] T. Müller, A. Evans, C. Schied, and A. Keller. "Instant Neural Graphics Primitives with a Multiresolution Hash Encoding". In: *ACM Transactions on Graphics* 41.4 (2022), pp. 1–15.

[NT09] D. Needell and J. A. Tropp. "CoSaMP: Iterative Signal Recovery from Incomplete and Inaccurate Samples". In: *Applied and Computational Harmonic Analysis* 26.3 (2009), pp. 301–321.

[Nem+09] A. Nemirovski, A. Juditsky, G. Lan, and A. Shapiro. "Robust Stochastic Approximation Approach to Stochastic Programming". In: *SIAM Journal on Optimization* 19.4 (2009), pp. 1574–1609.

[ND21] A. Q. Nichol and P. Dhariwal. "Improved Denoising Diffusion Probabilistic Models". In: *International Conference on Machine Learning*. 2021.

[OF97] B. A. Olshausen and D. J. Field. "Sparse Coding with an Overcomplete Basis Set: A Strategy Employed by V1?" In: *Vision Research* 37.23 (1997), pp. 3311–3325.

[OS20] S. Oymak and M. Soltanolkotabi. "Towards Moderate Overparameterization: Global Convergence Guarantees for Training Shallow Neural Networks". In: *IEEE Journal on Selected Areas in Information Theory* (2020).

[PR17] T. Plotz and S. Roth. "Benchmarking Denoising Algorithms With Real Photographs". In: *Conference on Computer Vision and Pattern Recognition*. 2017, pp. 1586–1595.

[Poo+23] B. Poole, A. Jain, J. T. Barron, and B. Mildenhall. "DreamFusion: Text-to-3D Using 2D Diffusion". In: *International Conference for Learning Representations*. 2023.

[Rad+21] A. Radford et al. "Learning Transferable Visual Models From Natural Language Supervision". In: *International Conference on Machine Learning*. 2021.

[RMC16] A. Radford, L. Metz, and S. Chintala. "Unsupervised Representation Learning with Deep Convolutional Generative Adversarial Networks". In: *International Conference for Learning Representations*. 2016.

[RSS12] A. Rakhlin, O. Shamir, and K. Sridharan. "Making Gradient Descent Optimal for Strongly Convex Stochastic Optimization". In: *International Conference on Machine Learning*. 2012.

[Rec+19] B. Recht, R. Roelofs, L. Schmidt, and V. Shankar. "Do ImageNet Classifiers Generalize to ImageNet?" In: *International Conference on Machine Learning*. 2019.

[RS19] E. T. Reehorst and P. Schniter. "Regularization by Denoising: Clarifications and New Interpretations". In: *IEEE Transactions on Computational Imaging* 5.1 (2019), pp. 52–67.

[RMW14] D. J. Rezende, S. Mohamed, and D. Wierstra. "Stochastic Backpropagation and Approximate Inference in Deep Generative Models". In: *International Conference on Machine Learning*. 2014.

[REM17] Y. Romano, M. Elad, and P. Milanfar. "The Little Engine That Could: Regularization by Denoising (RED)". In: *SIAM Journal on Imaging Sciences* 10.4 (2017), pp. 1804–1844.

[RFB15] O. Ronneberger, P. Fischer, and T. Brox. "U-Net: Convolutional Networks for Biomedical Image Segmentation". In: *Medical Image Computing and Computer-Assisted Intervention*. 2015.

[Rot09] L. Roth. "Looking at Shirley, the Ultimate Norm: Colour Balance, Image Technologies, and Cognitive Equity". In: *Canadian Journal of Communication* 34.1 (2009), pp. 111–136.

[RB09] S. Roth and M. J. Black. "Fields of Experts". In: *International Journal of Computer Vision* 82.2 (2009), p. 205.

[RV10] M. Rudelson and R. Vershynin. "Non-Asymptotic Theory of Random Matrices: Extreme Singular Values". In: *Proceedings of the International Congress of Mathematicians*. Vol. 3. 2010, pp. 1576–1602.

[ROF92] L. I. Rudin, S. Osher, and E. Fatemi. "Nonlinear Total Variation Based Noise Removal Algorithms". In: *Physica D: Nonlinear Phenomena* 60.1 (1992), pp. 259–268.

[SK16] T. Salimans and D. P. Kingma. "Weight Normalization: A Simple Reparameterization to Accelerate Training of Deep Neural Networks". In: *Neural Information Processing Systems*. 2016.

[Sch+18] J. Schlemper, J. Caballero, J. V. Hajnal, A. N. Price, and D. Rueckert. "A Deep Cascade of Convolutional Neural Networks for Dynamic MR Image Reconstruction". In: *IEEE Transactions on Medical Imaging* 37.2 (2018), pp. 491–503.

[Sha19] O. Shamir. "Exponential Convergence Time of Gradient Descent for One-Dimensional Deep Linear Neural Networks". In: *Conference on Learning Theory*. 2019.

[SB06] H. Sheikh and A. Bovik. "Image Information and Visual Quality". In: *IEEE Transactions on Image Processing* 15.2 (2006), pp. 430–444.

[Shi+22] E. Shimron, J. I. Tamir, K. Wang, and M. Lustig. "Implicit Data Crimes: Machine Learning Bias Arising from Misuse of Public Data". In: *Proceedings of the National Academy of Sciences* 119.13 (2022).

[SO01] E. P. Simoncelli and B. A. Olshausen. "Natural Image Statistics and Neural Representation". In: *Annual Review of Neuroscience* 24.1 (2001), pp. 1193–1216.

[Sit+20] V. Sitzmann, J. Martel, A. Bergman, D. Lindell, and G. Wetzstein. "Implicit Neural Representations with Periodic Activation Functions". In: *Neural Information Processing Systems*. 2020.

[Soh+15] J. Sohl-Dickstein, E. Weiss, N. Maheswaranathan, and S. Ganguli. "Deep Unsupervised Learning Using Nonequilibrium Thermodynamics". In: *International Conference on Machine Learning*. 2015.

[Son+22] J. Song, A. Vahdat, M. Mardani, and J. Kautz. "Pseudoinverse-Guided Diffusion Models for Inverse Problems". In: *International Conference on Learning Representations*. 2022.

[SE19] Y. Song and S. Ermon. "Generative Modeling by Estimating Gradients of the Data Distribution". In: *Neural Information Processing Systems*. 2019.

[SE20] Y. Song and S. Ermon. "Improved Techniques for Training Score-Based Generative Models". In: *Neural Information Processing Systems*. 2020.

[Son+21] Y. Song, J. Sohl-Dickstein, D. P. Kingma, A. Kumar, S. Ermon, and B. Poole. "Score-Based Generative Modeling through Stochastic Differential Equations". In: *International Conference for Learning Representations*. 2021.

[Sri+20] A. Sriram, J. Zbontar, T. Murrell, A. Defazio, C. L. Zitnick, N. Yakubova, F. Knoll, and P. Johnson. "End-to-End Variational Networks for Accelerated MRI Reconstruction". In: *Medical Image Computing and Computer Assisted Intervention*. 2020.

[Ste81] C. M. Stein. "Estimation of the Mean of a Multivariate Normal Distribution". In: *The Annals of Statistics* 9.6 (1981), pp. 1135–1151.

[Sze+14] C. Szegedy, W. Zaremba, I. Sutskever, J. Bruna, D. Erhan, I. Goodfellow, and R. Fergus. "Intriguing Properties of Neural Networks". In: *International Conference on Learning Representations*. 2014.

[Tan+20] M. Tancik, P. Srinivasan, B. Mildenhall, S. Fridovich-Keil, N. Raghavan, U. Singhal, R. Ramamoorthi, J. Barron, and R. Ng. "Fourier Features Let Networks Learn High Frequency Functions in Low Dimensional Domains". In: *Neural Information Processing Systems*. 2020.

[Tik77] A.-I. N. Tikhonov. *Solutions of Ill Posed Problems*. Washington: New York: John Wiley & Sons Inc, 1977.

[Tro04] J. A. Tropp. "Greed Is Good: Algorithmic Results for Sparse Approximation". In: *IEEE Transactions on Information Theory* 50.10 (2004), pp. 2231–2242.

[TG07] J. A. Tropp and A. C. Gilbert. "Signal Recovery From Random Measurements Via Orthogonal Matching Pursuit". In: *IEEE Transactions on Information Theory* 53.12 (2007), pp. 4655–4666.

[UVL18] D. Ulyanov, A. Vedaldi, and V. Lempitsky. "Deep Image Prior". In: *Conference on Computer Vision and Pattern Recognition*. 2018.

[VV+18] D. Van Veen, A. Jalal, E. Price, S. Vishwanath, and A. G. Dimakis. "Compressed Sensing with Deep Image Prior and Learned Regularization". In: *arXiv:1806.06438 [cs, math, stat]* (2018).

[Vas+17] A. Vaswani, N. Shazeer, N. Parmar, J. Uszkoreit, L. Jones, A. N. Gomez, L. Kaiser, and I. Polosukhin. "Attention Is All You Need". In: *Neural Information Processing Systems*. 2017.

[Vem05] S. S. Vempala. *The Random Projection Method*. American Mathematical Society, 2005.

[VBW13] S. V. Venkatakrishnan, C. A. Bouman, and B. Wohlberg. "Plug-and-Play Priors for Model Based Reconstruction". In: *IEEE Global Conference on Signal and Information Processing*. 2013.

[Vin11] P. Vincent. "A Connection Between Score Matching and Denoising Autoencoders". In: *Neural Computation* 23.7 (2011), pp. 1661–1674.

[Wan+20] F. Wang, Y. Bian, H. Wang, M. Lyu, G. Pedrini, W. Osten, G. Barbastathis, and G. Situ. "Phase Imaging with an Untrained Neural Network". In: *Light: Science & Applications* 9.1 (2020), pp. 1–7.

[WSB03] Z. Wang, E. Simoncelli, and A. Bovik. "Multiscale Structural Similarity for Image Quality Assessment". In: *Asilomar Conference on Signals, Systems Computers*. Vol. 2. 2003, 1398–1402 Vol.2.

[WB09] Z. Wang and A. C. Bovik. "Mean Squared Error: Love It or Leave It? A New Look at Signal Fidelity Measures". In: *IEEE Signal Processing Magazine* 26.1 (2009), pp. 98–117.

[Wil+19] F. Williams, T. Schneider, C. Silva, D. Zorin, J. Bruna, and D. Panozzo. "Deep Geometric Prior for Surface Reconstruction". In: *Conference on Computer Vision and Pattern Recognition*. 2019.

[WR22] S. J. Wright and B. Recht. *Optimization for Data Analysis*. Cambridge University Press, 2022.

[Wu+24] R. Wu et al. "ReconFusion: 3D Reconstruction with Diffusion Priors". In: *Conference on Computer Vision and Pattern Recognition*. 2024.

[Yam+20] B. Yaman, S. A. H. Hosseini, S. Moeller, J. Ellermann, K. Uurbil, and M. Akakaya. "Self-Supervised Learning of Physics-Guided Reconstruction Neural Networks without Fully Sampled Reference Data". In: *Magnetic Resonance in Medicine* 84.6 (2020), pp. 3172–3191.

[Yoo+21] J. Yoo, K. H. Jin, H. Gupta, J. Yerly, M. Stuber, and M. Unser. "Time-Dependent Deep Image Prior for Dynamic MRI". In: *IEEE Transactions on Medical Imaging* 40.12 (2021), pp. 3337–3348.

[ZMH15] M. Zaitsev, J. Maclaren, and M. Herbst. "Motion Artifacts in MRI: A Complex Problem with Many Partial Solutions". In: *Journal of Magnetic Resonance Imaging* 42.4 (2015), pp. 887–901.

[Zam+22] S. W. Zamir, A. Arora, S. Khan, M. Hayat, F. S. Khan, and M.-H. Yang. "Restormer: Efficient Transformer for High-Resolution Image Restoration". In: *Conference on Computer Vision and Pattern Recognition*. 2022.

[Zbo+19] J. Zbontar et al. *fastMRI: An Open Dataset and Benchmarks for Accelerated MRI*. 2019. arXiv: 1811.08839.

[Zha+17] K. Zhang, W. Zuo, Y. Chen, D. Meng, and L. Zhang. "Beyond a Gaussian Denoiser: Residual Learning of Deep CNN for Image Denoising". In: *IEEE Transactions on Image Processing* 26.7 (2017), pp. 3142–3155.

[Zha+18] R. Zhang, P. Isola, A. A. Efros, E. Shechtman, and O. Wang. "The Unreasonable Effectiveness of Deep Features as a Perceptual Metric". In: *Conference on Computer Vision and Pattern Recognition*. 2018.

[Zha+22] R. Zhao, B. Yaman, Y. Zhang, R. Stewart, A. Dixon, F. Knoll, Z. Huang, Y. W. Lui, M. S. Hansen, and M. P. Lungren. "fastMRI+, Clinical Pathology Annotations for Knee and Brain Fully Sampled Magnetic Resonance Imaging Data". In: *Scientific Data* 9.1 (2022), p. 152.

[Zho+21] E. D. Zhong, T. Bepler, B. Berger, and J. H. Davis. "CryoDRGN: Reconstruction of Heterogeneous Cryo-EM Structures Using Neural Networks". In: *Nature Methods* 18.2 (2021), pp. 176–185.

[Zho+23] B. Zhou, N. Dey, J. Schlemper, S. S. Mohseni Salehi, C. Liu, J. S. Duncan, and M. Sofka. "DSFormer: A Dual-domain Self-supervised Transformer for Accelerated Multi-contrast MRI Reconstruction". In: *IEEE/CVF Winter Conference on Applications of Computer Vision*. 2023.

[Zho+04] Zhou Wang, A. C. Bovik, H. R. Sheikh, and E. P. Simoncelli. "Image Quality Assessment: From Error Visibility to Structural Similarity". In: *IEEE Transactions on Image Processing* 13.4 (2004), pp. 600–612.

Index